Robert Theodor Wilhelm Lüpke

The elements of electro-chemistry treated experimentally

Robert Theodor Wilhelm Lüpke

The elements of electro-chemistry treated experimentally

ISBN/EAN: 9783337275525

Printed in Europe, USA, Canada, Australia, Japan

Cover: Foto ©berggeist007 / pixelio.de

More available books at **www.hansebooks.com**

THE ELEMENTS

OF

ELECTRO-CHEMISTRY

TREATED EXPERIMENTALLY

BY

DR. ROBERT LÜPKE

*Headmaster of the Municipal Dorothea Realgymnasium, and Lecturer
in the Imperial School of Posts and Telegraphs, Berlin*

TRANSLATED FROM THE SECOND, REVISED AND ENLARGED, EDITION BY

M. M. PATTISON MUIR, M.A.

Fellow and Lecturer of Gonville and Caius College, Cambridge

With Fifty-four Figures in the Text

LONDON
H. GREVEL & CO.
PHILADELPHIA: J. B. LIPPINCOTT COMPANY
1897

PREFACE

TO THE FIRST EDITION.

THE researches in physical chemistry which have been prosecuted with great eagerness during the last two decades have led to results whereby many of the problems of exact science which had remained unsolved have been answered. For instance, a deeper insight has been obtained into the nature of solutions, by the work of van't Hoff, who has shown that the law of Avogadro, which had been applied only to gases, holds good also for substances in solution. This result has been of especial importance to electro-chemistry, although at first sight it appears to have no connection with that subject. It may be affirmed to-day that the conduction of the galvanic current in electrolytes, as well as the origin of the current in galvanic cells, matters which had been discussed for a century, have been explained.

The science of electro-chemistry is set forth in detail in researches published in various journals, and in the text-books of physical chemistry of Ostwald and Nernst. Nevertheless I have undertaken to write this short book, because I thought the time to be opportune for bringing together the new results in a condensed form, and for giving a short survey to those who are not in a position to make an exhaustive study of the voluminous literature of the subject for themselves. In order to make the theories, which are difficult in themselves, more easily intelligible to the reader, who is assumed not to have more than a knowledge of the fundamental conceptions of physics and chemistry, the various laws have been deduced directly from the experimental results. Mathematical discussions have been introduced for the purpose of more complete confirmation only in a very few cases.

The experiments which form an essential part of this book are carried out with the simplest possible apparatus.* They are arranged to make the process that is under consideration clearly intelligible, and also to ensure the attainment of the final result in as short a time as possible. For this reason they may be made especially useful for teaching, and also for becoming practically acquainted with the physical and chemical properties of various substances. It is true that very accurate results are not to be expected in the experiments wherein measurements are made; for more delicate instruments, such as are to be found only in scientific laboratories, are required in order to obtain accurate numerical results; but, at the same time,

* Much of the apparatus may be prepared by the student himself. The rest can be obtained from Glass-blower Max Stuhl (Berlin, N., Philippstrasse 22).

such instruments are not generally suitable for the purposes of demonstration.

Although the main purpose of the book is to set forth the purely scientific aspects of electro-chemistry, the practical sides of the subject have not been left altogether unnoticed. Technical electro-chemical processes, and especially the processes of electro-metallurgy, which are so important at present, are referred to in their proper places.

ROBERT LÜPKE.

BERLIN, *May 1st,* 1895.

PREFACE

TO THE SECOND EDITION.

A SECOND Edition of my "Elements of Electro-chemistry, treated experimentally," was called for a few months after the appearance of the book. This showed me that there was want of a book setting forth succinctly the most important parts of electro-chemistry, and that the method of the book, which deduces the laws and elucidates the theories from the basis of experiments, had met with approval. Several works on electro-chemistry, it is true, have appeared in the course of a year, such as those of Ostwald, Jahn, Le Blanc, and Ahrens, to which my "Elements" are much inferior, looked at from the scientific standpoint. Nevertheless I hope that those readers who wish to make a preliminary survey of electro-chemistry before they enter on the study of the more detailed works will receive the second edition of my book with favour.

So far as time allowed, I have enlarged the first edition here and there, and made it more complete. The chapter dealing with the energetics of the galvanic elements has been re-written; and more attention has been given to the technical side of electro-chemistry, as regards the principles of the processes that have received practical applications.

ROBERT LÜPKE.

BERLIN, *April 18th*, 1896.

TRANSLATOR'S NOTE.

A FEW small changes have been made in the English edition, with the sanction of the Author.

The very small additions made here and there by the Translator are included in square brackets.

March, 1897.

TABLE OF CONTENTS.

	PAGES
INTRODUCTION	1–2

PART I.
RECENT THEORIES OF ELECTROLYSIS.

CHAPTER I. THE PHENOMENA OF ELECTROLYSIS . . . 4–31

(i.) *The Electrolysis of fused compounds* 4–9
    Preparation of magnesium from magnesium-potassium chloride   4
        „   „ aluminium from aluminium-potassium chloride .   6
    Calcium carbide and silicon carbide 8
    Electrolysis of fused lead chloride 8
        „   „ caustic potash 9

(ii.) *The Electrolysis of compounds in solution* 9–19
    Electrolysis of zinc chloride 9
        „   „ stannous chloride 10
        „   „ hydrochloric acid 11
        „   „ sodium chloride, and electrical bleaching . . 12
    Electrical preparation of potassium chlorate 13
    Electrolysis of copper sulphate between copper electrodes . 14
    Principle of electro-plating and etching 15–16
    Electrolysis of copper sulphate between a copper kathode and a platinum anode 16
    Electrolysis of dilute sulphuric acid between a platinum kathode and a copper anode 17
    Electrolysis of alkali salts of oxyacids 18
    Reagents for discriminating the galvanic poles . . . 19
    The conceptions of Berzelius and Daniell regarding salts . 20
    Hittorf's conception of an electrolyte 22

TABLE OF CONTENTS.

	PAGES
Electrolysis of dilute sulphuric acid	22
Electrolytic preparation of ozone	23
„ „ „ nitrogen chloride	24
Electrolysis of ammonia	24
Colouring metals	25
Electro-silvering and gilding	26
Separation of gold by the potassium cyanide process	27
Electrolysis of potassium ferrocyanide	27
„ „ sodium acetate	28
Electrical preparation of organic compounds	29–31

CHAPTER II. FARADAY'S LAW 32–40
   Deduction of the law from experimental results . . 32–34
   The electro-chemical equivalent 35
   Measurement of quantity of current by the voltameter . 36
   The laws of the subdivision of the current . . . 36–37
   Helmholtz's theory of the conduction of the current in electrolytes 37–39
   The absolute valency-charges of the ions . . . 40

CHAPTER III. HITTORF'S TRANSPORT-NUMBERS . 41–44

CHAPTER IV. THE LAW OF KOHLRAUSCH . . 45–56
   Measurement of the resistances of electrolytes . . 45
   Conductivities of electrolytes . . . 46–48
   The law of molecular conductivity . . . 49
   Velocities of migration of the ions . . . 50–52
   Ostwald's expression for the molecular conductivity . 51
   Demonstration of the migration-velocities of the ions . 52–56

CHAPTER V. THE DISSOCIATION THEORY OF ARRHENIUS . 57–73
   Work done by the current during electrolysis . 57–58
   Coefficient of dissociation 59
   Conductivity of pure water 60
   Capability of dissociation of the solvent . . 60–62
   Heat of ionisation according to Ostwald . . 63–66
   Theories of Grotthus and Clausius regarding current-conduction 67–68
   Energy-contents of the ions 68
   Explanation of chemical change by the dissociation theory; activity-coefficients 69–70
   Thermoneutrality of solutions 70
   Proof of the existence of free ions . . . 71–73

PART II.
THE THEORY OF SOLUTIONS OF VAN'T HOFF.

	PAGES
CHAPTER I. OSMOTIC PRESSURE	75–91
Diffusion of dissolved substances and the conception of osmotic pressure	75–76
Semipermeable membranes	77
Plasmolysis and the law of H. de Vries	77
Traube's proof of osmotic pressure	78
Pfeffer's measurement of osmotic pressure	79–83
„ laws of osmotic pressure	84
Horstmann's gaseous equation	84–87
The law of van't Hoff	88
Osmotic work done by solutions	89–90
Magnitudes of osmotic pressures	91
CHAPTER II. THE VAPOUR-PRESSURES OF SOLUTIONS	92–99
Vapour-pressure of a solution	92
Raoult's laws of vapour-pressures of solutions	93
„ measurement of vapour-pressures of solutions	94–97
Relation between osmotic pressure and vapour-pressure of a solution (Ostwald)	97–99
CHAPTER III. BOILING POINTS AND FREEZING POINTS OF SOLUTIONS	100–110
Rise of boiling, and lowering of freezing, point of a solution	100–101
Beckmann's measurement of boiling point	101–104
Raoult's laws of increment of boiling point	104
Determination of molecular weights by the boiling point method	105–106
Raoult's laws of lowering of freezing point	107–108
Determination of molecular weights by the freezing point method	109–110
CHAPTER IV. SUMMARY	111–116
Relations between osmotic pressure and decrease of vapour-pressure, increase of boiling point, and lowering of freezing point (van't Hoff)	111–113
Calculation of molecular lowering of freezing point by van't Hoff	113–116
CHAPTER V. AQUEOUS SOLUTIONS OF ELECTROLYTES	117–121
Carrying over of the law of van't Hoff to solutions of electrolytes	117–118
Explanation of the factor i by Arrhenius, and confirmation of the dissociation theory	119–121

PART III.

THE OSMOTIC THEORY OF THE CURRENT OF GALVANIC CELLS.

	PAGES
CHAPTER I. LIQUID CELLS	123–127
Calculation of E.M.F. of these cells by Nernst	124–126
CHAPTER II. CONCENTRATION-CELLS	128–135
Direction of the current in galvanic cells	129
Conception of electrolytic solution-pressure	129
Equations for the potential-differences of concentration-cells	130–132
Recognition of the current of concentration-cells	133
Tin tree	134
Mercurous nitrate cell	135
CHAPTER III. DANIELL CELLS	136–145
Equations for the potential-differences of Daniell cells	136–137
Measurement of E.M.F. by the compensation method	137–139
Experimental confirmation by Nernst's equation	140–142
Amalgam cells	142
Analogy between the galvanic current and the conduction of water	143–145
CHAPTER IV. REDUCTION-CELLS AND OXIDATION-CELLS	146–156
Chemical changes in Daniell cells	146–147
Recognition of the currents of various reduction- and oxidation-cells	148–150
Cell in which silver chloride is precipitated	150–151
Cells wherein acids are neutralised	151–152
Gas-cells (gas-accumulators)	152–156
Ostwald's element of the future	156
CHAPTER V. THE SOLUTION-PRESSURES OF THE METALS	157–170
Dropping electrodes	157–159
Potential-differences between metals and solutions of their salts	160–161
Calculation of solution-pressures of metals	162–164
Determination of E.M.F. of Daniell cells from solution-pressures	164
Electromotive series	165
Electromotive position of hydrogen	166
Influence of atmospheric air on combinations of metals	167–168
Rusting of galvanised and tinned iron	169–170

TABLE OF CONTENTS.

	PAGES
CHAPTER VI. INTENSITY OF FIXATION, AND POLARISATION	171–185
Electrolysis with soluble and insoluble anodes	171–172
Polarisation-currents	173–175
Le Blanc's intensity of fixation of the ions	176
Relation of E.M.F. of Daniell cells to intensities of fixation	177
Intensities of fixation of potassium and hydrogen ions	178
Decomposition-tensions of electrolytes	179
Electrolytic separation of the metals	181–183
Electrical purification of copper	183
Electrical preparation of copper by the processes of Siemens and Höpfner	184
CHAPTER VII. IRREVERSIBLE CELLS	186–196
Polarisation of inconstant cells	186
Constant irreversible cells	188
Depolarisation in Bunsen cells	189
Actions of different depolarising agents	192
Leclanché cells	193
CHAPTER VIII. ACCUMULATORS	197–206
The action of the electrodes in leaden accumulators	197
Planté's method of preparing accumulators	200
Charging accumulators by means of copper elements	201
The zero-effect of accumulators	204
Self-discharging accumulators	205
CHAPTER IX. THE ENERGETICS OF GALVANIC ELEMENTS	207–218
Total heat in the element and in the connecting wire	207
Chemical work done by the current	211
Transformation of chemical energy into electrical energy	212
Chemical changes in leaden accumulators	216
Peltier's effect and its relation to the temperature-coefficients of galvanic elements	217
INDEX	219

THE ELEMENTS OF
ELECTRO-CHEMISTRY.

INTRODUCTION.

A HUNDRED years have passed since the discovery of the voltaic battery, that simple apparatus which has been the starting-point of all those instruments wherein galvanic currents are obtained by combining conductors of the first and second class. But although voltaic batteries have been used for so long a time it was not until quite recently that the theory of their action was made clear; for the theories that had been advanced, the contact theory and the chemical theory, did not suffice to give a satisfactory explanation of the production of the galvanic current.

Nernst was the first to give a clear representation of the mechanism of the formation of the current by his *osmotic theory*, put forward by him at Göttingen in 1889.

The theory of Nernst is based on a broad and firm foundation. It assumes certain other theories, also of recent date, which are concerned with the most important parts of the youngest branch of scientific chemistry, to wit physical chemistry. It takes especial account of Helmholtz's theory of current-conduction, Arrhenius' theory of electrolytic dissociation [or ionisation], and van't Hoff's theory of solutions. These theories have all found their proper acknowledgment everywhere at the present time, not only because they rest on sufficient empirical observation, but also because they have most thoroughly explained many physical and chemical phenomena which had before been obscure.

These theories will be dealt with in the first and second parts of this book, and Nernst's theory of current-production will be considered in the third part.

PART I.

RECENT THEORIES OF ELECTROLYSIS.

THE term *electrolysis* refers to the chemical changes which occur when an electric current passes through a *conductor of the second class*. Such a conductor, which is always a chemical compound either in solution or fused, is called an *electrolyte* (λύω = loose). *Conductors of the first class*, which, in the form of wires or plates, lead the current into, or away from, electrolytic cells, are called *electrodes* (ὁδόω = lead in), one being called the *anode* and the other the *kathode* (ἡ ἄνοδος = the way up; ἡ κάθοδος = the way down). The two latter terms are derived from the supposition that the earth's magnetism is caused by electrical currents which flow around the earth parallel to the degrees of latitude, in the direction from east to west, that is from the rising to the setting sun. Because of the difference of potential that is caused on the electrodes by the electrolysing current, the two essentially distinct particles of the electrolyte move in different directions, one towards the anode and the other towards the kathode. These minute particles, which travel in the directions just indicated, are called *ions* (ἰών, gen. ἰόντος, = going); *anions* being those which go to the anode, and *kations* those which go to the kathode. The chemical changes brought about by the ions at the electrodes are called electrolysis. As we shall see in the first chapter, the results of these changes may differ very much, according to the conditions that prevail.

CHAPTER I.

THE PHENOMENA OF ELECTROLYSIS.

THE' results of electrolysis are shown in the simplest way when *fused* binary compounds are used, because in these cases the two ions separate directly on the electrodes.

Gorup-Besanez (*Anorganische Chemie*, 1871, p. 517) has arranged a lecture-experiment, with a clay tobacco-pipe, for demonstrating the electrolysis of magnesium chloride, which was first performed by Bunsen. Potassium-magnesium chloride, a salt that can be fused easily in a platinum dish, is used as the electrolyte. The salt is obtained in the fused condition by evaporating to dryness, on the water-bath in a platinum basin, a solution of 20 grams crystallised magnesium chloride and 7·5 grams potassium chloride, with addition of 3 grams ammonium chloride, and then rapidly heating the solid residue by a blowpipe-flame. The fused mass is poured into the bowl of an unglazed pipe (p in fig. 1), which has been previously strongly heated, and which is supported in a clamp; the current from ten accumulators * arranged *in series*

---

* The accumulators made by W. A. Böse & Co. (Berlin S.O. Köpenickerstrasse 154) are very suitable for laboratory purposes. The plates are not grated; they consist wholly of a mass of active material enclosed in a frame of hard lead. I employ cells 22 centims. high, 12 centims. wide, and 11 centims. long [about 9 × 5 × 4½ inches]. When discharged continuously, with 1·8 volts, they give 50 ampère-hours, if discharged with 2 ampères, and 28 ampère-hours if a current of 5·6 ampères is taken. Their capacity is, therefore, very considerable; the null effect (*see* Chapter VIII., Part III.) amounts in ampère-hours to 91 per cent.

(that is, one behind the other) is then sent through the fused salt, a knitting-needle which is passed through the stem of the pipe serving as kathode (k in fig. 1), and a rod of carbon which dips into the bowl being used as anode (a in fig. 1). Although the heat produced by the current soon liquefies the salt, yet it is better to place a small flame beneath the bowl of the pipe.

Chlorine is easily detected by bringing a piece of moistened litmus paper near the anode [the paper is bleached by the chlorine]. But the greater part of the magnesium burns on the surface of the fused salt, which gradually fumes and sputters, and when the apparatus is allowed to cool the magnesium is disseminated through the mass in such fine powder that its silvery lustre cannot be observed.

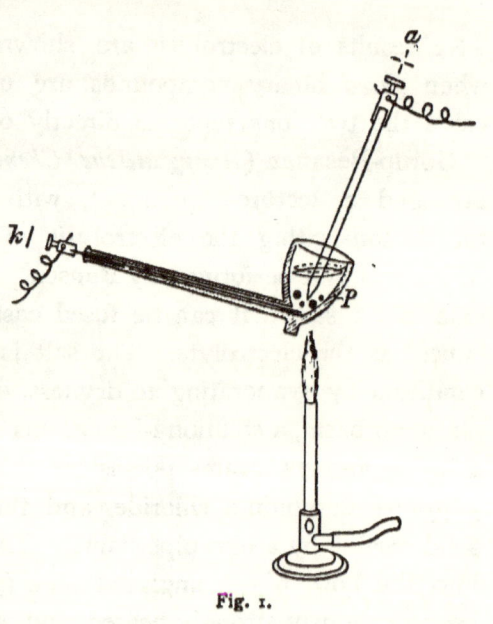

Fig. 1.

Both drawbacks are easily obviated by covering the fused electrolyte with a thick layer of finely powdered wood charcoal, immediately after closing the circuit. The sputtering is thus avoided, and after allowing the current to pass for barely twenty minutes a great many lustrous pellets of magnesium, from 1 to 2, and sometimes even 5, mm. diameter, can be obtained when the cold mass is broken up. The pellets should be isolated by rubbing the mass in a mortar along with alcohol; they may be burnt, with the production of bright light lasting from fifteen to thirty seconds, by wrapping them, singly, in a little loop of copper wire-gauze and igniting them in a flame. The pellets may be melted to a regulus by putting as much powdered fluorspar as will lie on the tip of a spatula into the bowl of the pipe, at the close of the electrolysis, and heating the bowl strongly for ten minutes.

Aluminium also can be obtained in the form of lustrous pellets by a process similar to that described; and the little particles may be united into one piece by throwing them into fusing common salt. The reaction characteristic of aluminium is obtained by dissolving the pellets in hydrochloric acid, adding an excess of pure potash solution, and precipitating aluminium hydroxide by addition of ammonium chloride.

There is however some difficulty in preparing anhydrous aluminium-potassium chloride, which was the salt used by Bunsen in 1854, and is the only salt suitable for employment as an electrolyte in this experiment. Dry aluminium chloride is prepared by passing a stream of dry hydrochloric acid gas over aluminium that is heated strongly. The easiest way of obtaining a current of hydrochloric acid gas, which will continue for four to six hours, and which can be regulated, is to allow concentrated sulphuric acid to drop into commercial hydrochloric acid. Figure 2 shows the arrangement of the apparatus; the hydrochloric acid is placed in the flask K, and the sulphuric acid drops from the stoppered funnel-tube H. The hydrochloric acid gas, dried by passing through concentrated sulphuric acid in the wash-bottles F_1 and F_2, flows into a tubulated vessel of about half a litre [say 17 oz.] capacity, which contains from five to ten grams of aluminium cut into small pieces, and which is heated by a large Bunsen lamp. After some time aluminium chloride is deposited in the wide neck of the receiver (h) as a white sublimate, and when the action has proceeded for two or three hours the deposit of chloride is removed by means of a knife. As aluminium chloride is very hygroscopic, the salt must be transformed into the more stable double chloride as soon as it is removed from the neck of

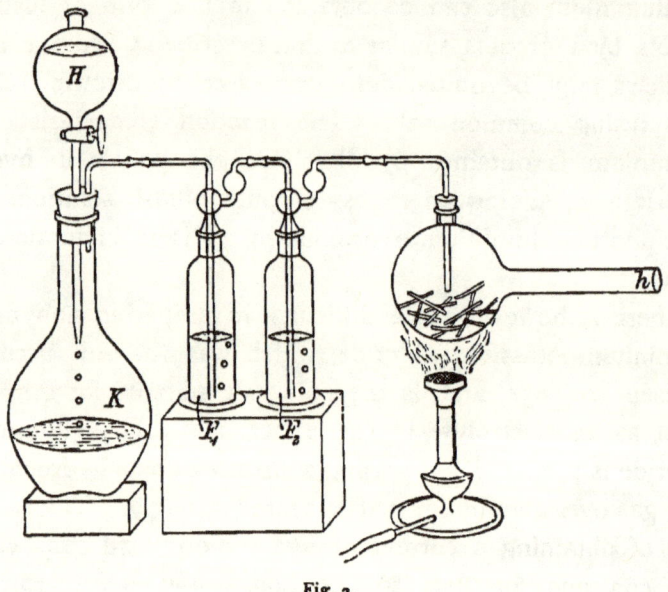

Fig. 2.

the receiver. For this purpose all that is required is to melt two parts of potassium chloride in a platinum crucible, and to add one part of aluminium chloride, in successive portions, stirring constantly, and then to pour out the melted mass on to a piece of porcelain. The double salt must be kept in a closely stoppered jar.

The principle of the experiment wherein the double aluminium-potassium chloride is electrolysed is the same as that of the technical preparation of aluminium by the electrolytic process. The main difference between the two processes lies in the fact that in the manufacturing operations pure aluminium oxide, prepared from the mineral *bauxite*, is used as the electrolyte; the oxide of aluminium is thrown into fusing *cryolite*, in a Héroult oven, and a current of great intensity is passed through the mass.

Special emphasis must be laid on the fact that in those preparations of very infusible metals which have been performed recently by electrical methods, for instance the preparation of chromium and tungsten (and also of the carbides of calcium and silicon*), the electric current does not cause electrolysis, but that it acts only as a source of heat which is sufficiently intense to cause the reduction of the oxides of the elements used, in the presence of carbon.

Fused lead chloride is much more easily decomposed by the current than compounds of magnesium or aluminium, and, for this reason, the lead salt is more suitable for demonstration purposes. As lead chloride is somewhat volatile when heated, the salt ought to be melted, in a porcelain crucible, in a draught place, before it is put into the bowl of the pipe. A sufficiently large regulus of lead is obtained by passing the current from five accumulators for about ten minutes. The melted regulus is poured out on to a piece of porcelain, and the metallic surface is exposed by rubbing with a file.

Potassium hydroxide is the most suitable base to electrolyse in the molten state. Sufficient mercury is poured into a platinum basin to cover the bottom of the dish; a few sticks of potash are placed in the basin, and these are melted by

---

* Calcium carbide is used in the production of acetylene gas, which is coming into demand for lighting purposes; while silicon carbide is used, under the name *Carborundum*, for polishing because of its very great hardness.

heating the vessel by a small flame; the current from about five accumulators is then sent through the melted potash, the basin being used as the kathode, while a piece of platinum foil sunk into the electrolyte serves as the anode. The decomposition proceeds in accordance with the scheme $2 KOH = 2 K + H_2O + O$. The potassium alloys with the mercury, and the oxygen passes off from the surface of the platinum foil. After about half an hour the liquid amalgam is poured into a test-tube, where it is allowed to solidify. It is easy to prove the presence of potassium by placing the alloy in a small flask arranged for preparing a gas, adding diluted hydrochloric acid, and collecting the hydrogen that comes off; about half a litre of hydrogen may be obtained in a short time.

As lead and chlorine can be separated from fused lead chloride, so the two constituents of zinc chloride, zinc and chlorine, can be obtained by the decomposing action of an electric current on a *concentrated aqueous solution* of the salt zinc chloride. The electro-

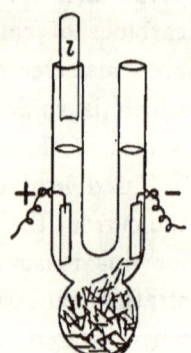

Fig. 3.

lysis is conducted in a bulbed U-tube, containing two platinum plates [connected with platinum wires which pass through the U-tube], the limbs of the tube being about the length and diameter of ordinary test-tubes. (*See* figure 3.) If ten accumulators are employed the bulb of the U-tube will be filled with elegant, branching crystals of zinc, after about twenty minutes; and if a piece of litmus paper is held in the limb of the tube which contains the anode, the paper will be bleached, very quickly, by the chlorine that is coming off.

If the anode is made of the metal of the salt which is electrolysed, the anode will be gradually dissolved by the anions that are separated in contact with it, and, at the same time, the metallic ions of the salt will be deposited on the kathode. Such a separation of metal is shown very prettily by electrolysing an aqueous solution of stannous chloride under the following conditions. The decomposition is effected in a large cylindrical vessel open at both ends (c in fig. 4), of from 1½ to 2 litres capacity, placed on a tripod; a cork is fitted into the bottom opening of the cylinder, and a copper wire passing through this cork is connected with the anode, a, which is a plate of cast tin about 7 centimetres broad. The kathode, k, is a flat-bottomed copper basin, supported from the cover closing the upper end of the cylinder by the wire for the conduction of the current, which wire is soldered to the basin. The electrolytic liquid is prepared by dissolving 65 grams of stannous chloride in warm hydrochloric acid, removing the excess of acid as far as possible by evaporation, and diluting with water to 1·5 litres. The intensity of the current is regulated so that hydrogen does not make its appearance on k. When the circuit is closed tin begins to separate in lustrous filaments, which grow visibly from the bottom of the basin k downwards through the liquid.

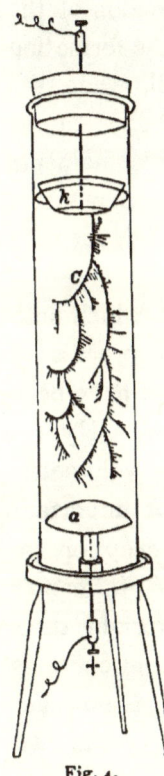

Fig. 4.

Figure 4 represents the appearance of the filaments of tin after the current has been passing for about twenty minutes. Branches grow at right angles to the primary filaments. At first these branches are equally large on both sides; but soon the growth of the branches takes place chiefly on one side, and as both the stems and their branches continue to increase in length, new branches appear, in a regular way, between those that have been formed already. Meanwhile the process of branching repeats itself on the branches which were first produced. One of these branches becomes larger than the others, and, as the increase of the upward-curving end of it becomes slower, this branch stretches in the direction of the stem, and the process begins anew. In this way the formation increases downwards in a regular manner, until, when the anode is nearly reached, the weight of the branches causes the stem to snap and the whole falls down. Other filaments here and there fall in the same way. Meanwhile new stems are formed, and these fill the upper third of the cylinder with their lustrous ramifications.

There are not many electrolytic solutions from which both ions can be separated on the electrodes by a process of electrolysis. The solution of stannous chloride we have been considering is an example of such an electrolyte. Hydrochloric acid also belongs to this category; the volumetric electrolysis of this compound is well known to be important in establishing the fundamental chemical conception of equivalency. This electrolysis is carried out, by Hofmann's method, in a U-tube, which is attached to an upright tube, and is furnished with a couple of glass stopcocks, and contains two electrodes of carbon (*compare* fig. 14, p. 33). It is however difficult to obtain exactly equal volumes of gas in the two limbs of the U-tube; this is due partly to the solubility of chlorine, and partly to the disengagement of oxygen from the anode, and the difficulty increases the more dilute the electrolytic liquid is made. After many trials, a satisfactory result was obtained by electrolysing, in a Hofmann's apparatus with carbon electrodes cut from pure gas-coke and provided with binding screws which were soldered on, a mixture of 10 c.c. pure hydrochloric acid of specific gravity 1·125 and 150 c.c.

very concentrated solution of calcium chloride of specific gravity 1·36. After allowing the current from five accumulators to pass for fifty minutes, the stopcocks being open, the liquid in the limb which contained the anode was quite saturated with chlorine; and when the stopcocks were then closed exactly 40 c.c. of chlorine were obtained for every 40 c.c. of hydrogen produced. It is necessary to guard against the possibility of a negative gaseous pressure by removing liquid from the upright tube, after closing the stopcocks in the two limbs, either by the use of a siphon, or, preferably, by means of a stopcock let into the lower part of the upright tube. The apparatus described by Lothar Meyer (*Berichte der d. Chem. Ges.*, **27**, 850 [1894]) depends on the same principle, and this apparatus has the advantage of employing pure hydrochloric acid as the liquid to be electrolysed.

Hydrogen and chlorine also appear at the electrodes in the electrolysis of a solution of common salt; the chlorine separates directly at the anode, and the hydrogen separates indirectly at the kathode, for each sodium ion turns out an atom of hydrogen from a molecule of water, so that soda is formed in the neighbourhood of the kathode ($Na + H_2O = NaOH + H$). Attempts have been eagerly made to turn this reaction to account in chemical technology, and so to obtain the products of the soda-industry in a simpler way. The most important part of this problem is the construction of a suitable diaphragm for keeping separate the substances that are produced at the two electrodes. Although this difficulty has not yet been completely overcome, the electrolysis of solutions of common salt has been conducted in several places on a large scale, more especially for bleaching purposes.

The electrical process of bleaching is easily understood from the following experiment. The four-sided vessel of a Böse accumulator (*see* fig. 51, and note p. 4) is used as a cell.

Three longitudinal grooves are cut in the thin side-walls of the vessel; a clay diaphragm is placed in the middle groove, and a plate cut from retort carbon, and provided with a binding screw, in each of the other grooves. A piece of red-dyed cotton is spread, by means of four little wooden pegs, on that carbon plate which is to be used as the anode. The electrolytic liquid, which is poured into the cell, is made by dissolving 50 grams of common salt, 5 grams of magnesium chloride, and a few drops of hydrochloric acid, in a litre of water. The cell is covered with a glass vessel, and the poles of a battery of five accumulators are placed on the electrodes; in a short time so much of the red cotton as is immersed in the liquid is completely bleached.

Chlorate of potassium is made in certain factories by the electrolysis of a solution of potassium chloride, the basic lye which is formed around the kathode being transported from time to time to the anode, where, at the higher temperature, the chlorine reacts in accordance with the equation

$$6KOH + 6Cl = KClO_3 + 5KCl + 3H_2O.$$

Most of the electrolytes that have been referred to are binary compounds, and in the majority of the cases mentioned the two constituents of the electrolyte separate directly at the electrodes. *Such a separation in two directions generally occurs as a consequence of electrolysis, however complicated the molecule of the electrolyte may be.* The hydrogen, or the metal, or a radicle which represents the hydrogen, is always carried towards the kathode, by the primary reaction, and the rest of the molecule travels towards the anode. Then, either the ions are set free on the electrodes, or secondary reactions occur thereat, between the ions and the substances of which the electrodes are made, or between the ions and the electrolyte or the water. The occurrence of these secondary reactions will be better understood by considering the following examples.

If a current of moderate intensity is sent through a concentrated solution of copper sulphate in a rectangular trough, copper electrodes being used, that part of the kathode which is immersed in the liquid is soon covered with a dull red deposit of copper, and the mass of the anode diminishes because each SO_4 ion dissolves an atom of copper. The result of the passage of the current is, therefore, merely to carry copper from the anode to the kathode.

The fact that the weight of the copper plate increases or decreases according as the plate is made the kathode or the anode may be demonstrated by Langley's experiment (*Zeitschrift f. physikal. Chemie*, 2, 83-91 [1888]), wherein the electrode in question is attached to one end of the beam of a balance in such a way that the swinging of the beam is not interfered with. Figure 5 represents a suitable arrangement. T is a funnel-shaped metallic vessel used as a counterpoise; the copper electrode, p_1, which is 8 centims. long by 3 centims. broad, is suspended by the platinum wire p on the hook h, and is completely immersed in the electrolyte (200 c.c. of a saturated solution of copper sulphate mixed with 15 c.c. of nitric acid) contained in the cell Z; the electrode hangs freely in the liquid so that it can move up and down without touching the walls of the cell. C is a commutator the binding screws of which are connected on the one hand with a battery, B, of two accumulators, and on the other hand with the screw s of the metal rod S, and with the plate of copper p_2, the size of which is 15 × 4 centims., and which is placed securely in the cell. The balance having been brought into equilibrium by placing small shot in the funnel T, the negative pole of the battery is connected with s; after about three minutes the beam inclines towards the side where the cell is placed.

CHAP. I.]   THE PHENOMENA OF ELECTROLYSIS.   15

The current is then reversed; after three minutes the beam again attains equilibrium, and after another three minutes it inclines in the opposite direction.

The experiment of electrolysing a solution of copper sulphate between copper electrodes illustrates the principle of electroplating, and also the principles of the electrical purifica-

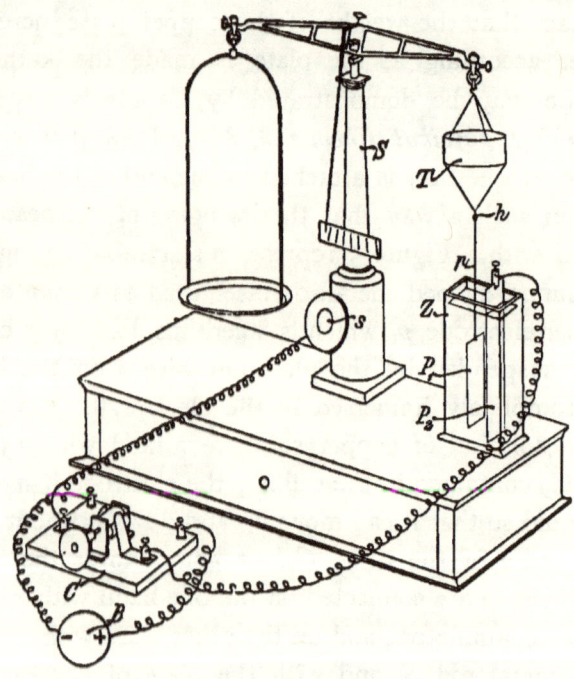

Fig. 5.

tion of crude copper and the electrical separation of copper from anodes of copper ore. (For more details see Part III., Chapter VI.) The experiment also elucidates the mechanical art of the galvanic etching of iron or copper vessels. The vessels are covered with a non-conducting etching varnish, and are then arranged as the anodes in acidified copper sulphate baths, after the pattern to be etched has been cut through the covering of varnish. The deeply etched parts of the vessels can then be filled with other metals (silver, gold, etc.) by placing the vessels, as they are taken from the copper baths, as kathodes in baths of the metals to be deposited; an imitation of Oriental inlaid metal-work is thus produced.

The kind of etching that has been described is easily carried out on a small scale. A plate of polished copper is covered with melted wax, by the use of a pencil; a pattern is then cut through the wax by a knitting-needle, and the plate is exposed for about twenty minutes as the anode [in an acidified bath of a copper salt] to the action of a current from four cells. When the wax is then removed by turpentine, the pattern is seen eaten into the plate.

If the copper anode in an acidified copper bath is replaced by a plate of platinum, oxygen is disengaged on the platinum, in accordance with the equation

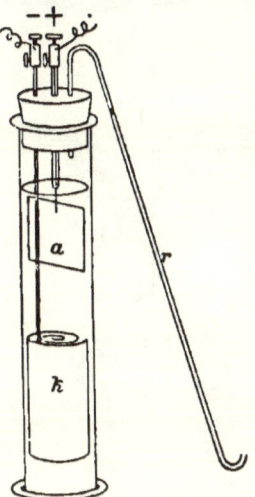

Fig. 6.

$$SO_4 + H_2O = H_2SO_4 + O;$$

for the chemical affinity of the platinum does not cause the formation of a sulphate under these experimental conditions. The electrolytic process is suitable for preparing small quantities of oxygen, the apparatus of Landolt, represented in figure 6, being used. The kathode, k, is a spirally rolled sheet of platinum; the strip of platinum which is soldered to k for conducting the current is covered with an isolating material. The platinum anode is marked a in the figure, and r is the tube for leading off the gas.

Figure 7 represents an apparatus used for demonstrating that hydrogen is set free on the platinum kathode, k, when a dilute solution of sulphuric acid is electrolysed, and that the ion SO_4 causes solution of copper from the copper anode, a, although copper does not otherwise dissolve in diluted sulphuric acid at the ordinary temperature. When a current from five accumulators has been passing for about ten

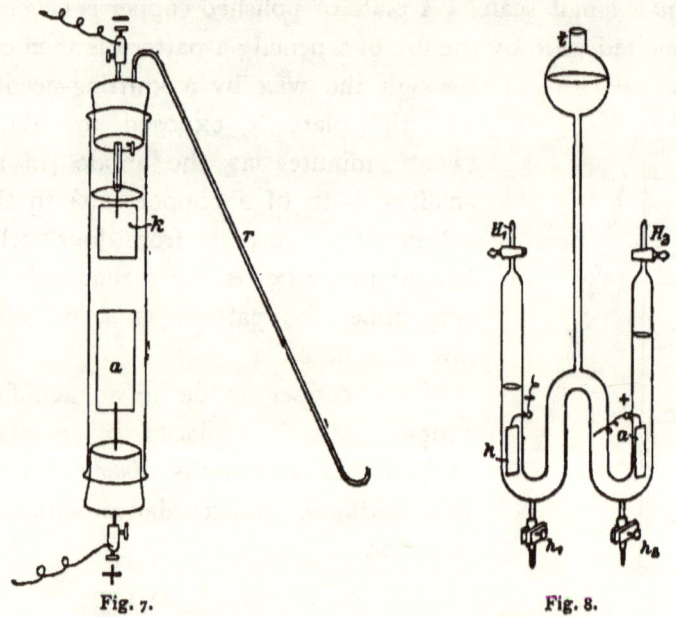

Fig. 7.   Fig. 8.

minutes, the liquid around the electrode a becomes distinctly blue, and pure hydrogen escapes by the tube r.

All acids and salts are decomposed by electrolysis, primarily, into hydrogen, or metal, and acid radicle.

The oxysalts of the alkali metals seem, at the first glance, to be exceptions to this rule. If a cold concentrated solution of potassium sulphate is electrolysed, in the apparatus represented by figure 8, between the platinum electrodes k and a, the current from ten accumulators being employed, two volumes of hydrogen collect in the limb containing the kathode, and one volume of oxygen forms in the limb which contains the anode; the gases may be recognised by opening the stopcocks H_1 and H_2. But the electrolyte has simultaneously undergone change at both electrodes: this may be shown by drawing off the liquid separately from each limb, by the stopcocks h_1 and h_2, and adding blue litmus to the liquid that surrounded the anode, and reddened litmus to the liquid which was around the kathode; the former liquid shows an acid, and the latter an alkaline, reaction.

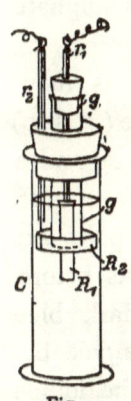

Fig. 9.

The experiment may be performed by pouring potassium sulphate solution containing a little blue litmus into the limb in which the anode is placed, and some of the same potassium sulphate solution containing a little reddened litmus into the limb where the kathode is fixed. Or the arrangement may be made still simpler by using an indicator which is affected differently by acids and alkalis: a solution of potassium sulphate (15 : 1,000), which is coloured deep red by an aqueous solution of cochineal, is placed in a Hofmann's U-tube (see fig. 14), and a current from three accumulators is sent through the solution; the liquid becomes violet around the kathode, and pale yellow around the anode.

The processes which take place at the electrodes during the electrolysis of alkali salts may be used for distinguishing between the poles of the source of a current. A small upright cylinder, C (fig. 9), is closed by a cork through which passes a glass tube, g; a wire passes through the tube r_1 and is connected with the platinum cylinder R_1; the wire passing through the tube r_2 leads to the ring of platinum foil R_2. The vessel is filled with a solution of an alkali salt (potassium sulphate or sodium chloride), to which a few drops of an alcoholic solution of phenolphthaleïn have been added. When the circuit is closed the liquid round one of the electrodes becomes coloured rose-red: that electrode is in connection with the negative pole of the battery which is being examined; for free alkali is formed at this electrode, and reacting with the phenolphthaleïn forms the red alkali salt of that [slightly acidic] compound. If the cylinder is shaken the red colour disappears, for the acid produced at the anode decomposes the salt and sets free phenolphthaleïn.*

The use of "*pole-reagent paper*" is based on these reactions. That reagent is blotting-paper soaked in potassium sulphate solution mixed with a very little solution of phenolphthaleïn; it must be moistened before use. The reaction is more delicate if the paper is impregnated with starch paste (2:100) to which has been added one part of potassium iodide and a very little phenolphthaleïn, then dried in air free from dust, and kept in a stoppered glass bottle. When the poles are laid on this paper the negative pole is recognised, as before, by the production of a red colour, and a very dark blue colour is produced on the paper under the positive pole because of the formation of "iodide of starch" by the iodine set free by the current. This paper enables an operator to recognise the sign of one of the two poles of a battery when the other is led to earth, as is often the case in telegraph offices; it is only necessary to lay the moistened paper on a conductor connected with the earth, and to touch it with the doubtful pole. The action of this reagent may be made clear to a larger audience by passing a current through the mixed solutions already mentioned, but diluted ten times,

* According to Ostwald (see *Foundations of Analytical Chemistry*, p. 118), the alkali salts of phenolphthaleïn are ionised in dilute solutions, and the red colour is caused by the free ions; when an acid is added non-ionised phenolphthaleïn is formed, and this is colourless.—[TR.]

in a U-tube furnished with platinum electrodes (*see* fig. 12, p. 27). The liquid in the anode limb very soon becomes black, and that in the kathode limb becomes red.

Because of the behaviour of the alkali oxysalts during electrolysis, Berzelius supposed that all salts contained a base and an acid (in the meaning then given to these terms) as their characteristic constituents; hence he wrote the formula of potassium sulphate $K_2O . SO_3$. He also supposed that the basic and the acidic oxides were exchanged, one for another, in chemical reactions between salts. The production of hydrogen and oxygen he regarded as due to a second, distinct, action of the current, namely the electrolysis of water. The fact that only copper, and no hydrogen, appears on the kathode during the electrolysis of copper sulphate between platinum electrodes was set down as the result of the reduction of copper oxide by the hydrogen coming from the water. This view was in keeping with his electro-chemical system; but it seemed that the chlorides must be looked on as exceptions, for they are decomposed by electrolysis directly into metal and chlorine. Neither was any explanation forthcoming of the fact that only an exchange of metals occurs when a haloid salt reacts with an oxysalt, whereas when oxysalts react they exchange bases.

These inconsistencies were removed by Daniell, at a later time. By including a water-voltameter in the circuit, besides the cell containing potassium sulphate solution, Daniell found that exactly the same quantities of hydrogen and oxygen were produced in the voltameter as were obtained during the electrolysis of the potassium sulphate. If, then, the view of Berzelius were correct a greater quantity of current must have gone through the cell than through the voltameter; but Faraday's law [see next chapter] shows that this is impossible.

Daniell gave an explanation of the electrolysis of salts which was free from objections. He said that the current decomposes potassium sulphate, as it decomposes every other salt, between platinum electrodes, into metal and acidic radicle: but in the case of potassium sulphate both these products of decomposition enter into secondary reactions with the water; the potassium reacts at the kathode in accordance with the equation

$$2K + 2H_2O = 2KOH + H_2,$$

and the SO_4 reacts at the anode in this way—

$$SO_4 + H_2O = H_2SO_4 + O.$$

The gases that are given off are, therefore, secondary products; and it is thus explained why the volumes of both gases are equal to the volumes of the same gases obtained in the voltameter, and are also equivalent to the quantities of acid and base observed at the electrodes. It is possible to separate potassium directly from solutions of its salts, because if mercury is used as the kathode the potassium forms an amalgam with that metal, and the action of water on this amalgam becomes apparent, by the evolution of little bubbles of hydrogen, only some time after the closing of the circuit.

In accordance with the results of electrolyses, Daniell said that *every salt is a compound of a metal, or a metal-like radicle, with an acidic residue*. The acidic residue is either a halogen or a group of different elements. Moreover, as the behaviour of hydrogen when heated (its conductivity), and its behaviour towards metals (its occlusion), show that hydrogen is to be looked on as a metal, and as the hydroxyl groups of the bases correspond with the acidic residues of salts, acids and bases may be included in the class of salts. Looking at the subject from this point of view, Hittorf made the following generalisation (*Über die Wanderungen der Ionen*, 2 Hälfte, p. 124 [*Ostwald's Klassiker*]): *Electrolytes are salts; they are separated by electrolysis into the same atoms, or atomic groups, as they exchange in their chemical reactions with one another.* All other substances, whether they be themselves liquids or whether they be dissolved, are non-conductors. This statement especially concerns the majority of the organic compounds, for it is only those organic compounds that have a salt-like character, in the meaning given to this term by Hittorf, which conduct the electric current.

Water is never primarily decomposed when a current is passed through aqueous solutions of electrolytes. Indeed, absolutely pure water does not conduct at all. The statement, which is made so often, that the sulphuric acid added to the water in a voltameter makes the water conduct, is only correct in so far that the acid separates, primarily, into H_2 and SO_4, and that the SO_4 ion is then reconverted into H_2SO_4 by reacting with the water and splitting off oxygen. One ought to speak of this reaction as an electrolysis of sulphuric acid. The addition to the water of some other oxyacid, or of a soluble base, or even of an alkali salt, makes it conduct, just as the addition of sulphuric acid enables the current to pass. The conductivity of naturally occurring water is to be attributed to the presence of alkali salts in it.

In conclusion, it is to be observed that the electrolysis of water acidified with sulphuric acid follows a somewhat different course when the acid is more concentrated and the intensity of the current is increased. If a mixture of one volume of sulphuric acid with five volumes of water is electrolysed, the oxygen which comes off at the anode is rich in ozone. The ozone is produced in accordance with the following reactions:

(1) $6HHSO_4 = 6H + 6HSO_4$
(2) $6HSO_4 + 3H_2O = 6H_2SO_4 + O_3.$

Considerable quantities of ozonised oxygen may be obtained by using the apparatus represented in figure 10. A good cork is fitted into a small cylinder; the cork carries the exit tube for gas, r, a narrower tube, g_1, and a wider tube, R. The wider tube, R, is closed by a cork through which pass the tube g_2 and also a short narrow tube. The tubes g_1 and g_2 are made of thermometer-tubing: the platinum wires a and k are attached to the lower ends of these tubes by means of fusible glass; these wires project only a short way out of the tubes, and they are soldered to pieces of copper wire which fill the bores of the tubes g_1 and g_2. When the positive pole of a battery of about five accumulators is connected with the wire in g_1, and the negative pole with the wire in g_2, hydrogen escapes by the narrow tube near g_2, and oxygen through r, and the presence of ozone in the oxygen is detected by the blue colour which is very quickly produced if the gas is caused to bubble through a solution of potassium iodide mixed with starch paste placed in the beaker-glass.

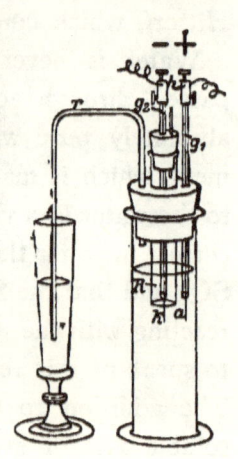

Fig. 10.

The quantity of ozone increases considerably if some chromium trioxide is added to the electrolytic liquid, and the apparatus is immersed in a freezing mixture.

This experimental arrangement may be used to show the production of hydrogen peroxide, besides ozone, during the electrolysis of dilute sulphuric acid. Hydrogen peroxide is always formed at the anode when the electrolyte contains at least 60 per cent. of H_2SO_4; the process may be represented by the equation

$$2HSO_4 + 2H_2O = H_2O_2 + 2H_2SO_4.$$

Considerable quantities of hydrogen peroxide are obtained by letting the current pass for half a minute through a mixture of 80 grams H_2SO_4 and 20 grams H_2O. A solution of 1 gram titanic acid in a hot mixture of 70 grams H_2SO_4 and 30 grams H_2O is used as a test for hydrogen peroxide; 3 c.c. of this solution are added to the electrolyte, and the peroxide is recognised by the deep yellow colour it produces.

That the ions can react with the electrolytes themselves may be proved by a great number of examples. The following cases will suffice to show how greatly the results of electrolytic processes may be varied by secondary reactions of this kind.

There is a well-known method for preparing nitrogen chloride by electrolysing a solution of salammoniac between a mercury kathode and a platinum anode (Heumann, *Anleitung zum Experimentiren*, p. 268 [1893]); the chlorine which is produced at the anode in that process reacts with undecomposed salammoniac in accordance with the equation

$$NH_4Cl + 6Cl = NCl_3 + 4HCl.$$

The electrolysis of a solution of ammonia is more complicated; the process may be represented by the scheme

$$\begin{array}{c|c} \overline{} & + \\ 6(NH_4) & 6(OH) \\ \hline 6NH_4 + 6H_2O = 6NH_4OH + 6H & 6(OH) = 3H_2O + 3O \\ & 2NH_4OH + 3O = 5H_2O + 2N. \end{array}$$

The radicle NH_4 interacts with water similarly to potassium, and six volumes of hydrogen are produced at the kathode. The oxygen which is separated from the 6(OH) ions oxidises two molecules of ammonium hydroxide to water and nitrogen; and thus it is that the volumes of hydrogen and nitrogen produced are in the ratio of 3 to 1. The experiment is best carried out in the apparatus used for the electrolysis of hydrochloric acid. Inasmuch as the nitrogen which is set free dissolves fairly easily in ammonia solution, it is best to employ as electrolyte a mixture of 250 c.c. concentrated solution of common salt and 20 c.c. concentrated ammonia solution, and to allow the current from six accumulators to pass through the apparatus for an hour before the stopcocks of the two limbs are closed and the stopcock of the upright tube is opened (*see* fig. 14, p. 33).

When an alkaline solution of a lead salt is electrolysed, the oxygen produced at the anode, as the result of secondary reactions, oxidises the lead compound to lead dioxide, as shown in the following equations:—

$$Pb(NO_3)_2 = Pb + 2NO_3,$$
$$2NO_3 + H_2O = 2HNO_3 + O,$$
$$Pb(NO_3)_2 + H_2O + O = 2HNO_3 + PbO_2.$$

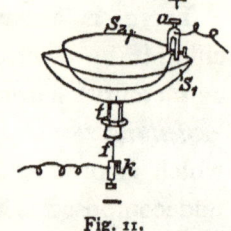

Fig. 11.

The experiment is easily performed. The glass basin S_1 (fig. 11), in the tubulus of which, t, is fastened the short iron wire f, is placed on a triangle, and is filled with a 5 per cent. solution of lead nitrate mixed with its own volume of a normal solution of caustic soda [a normal solution contains 40 grams of NaOH per litre]; a metallic plate, preferably a platinum basin, S_2, is immersed in the liquid in the glass vessel. When k has been connected with the kathode pole and a with the anode pole of an accumulator, for about fifteen seconds, four or five iridescent, rainbow-coloured rings of lead dioxide appear on the platinum basin, S_2. The art of colouring metals, which aims at ornamenting vessels of copper or brass that have been thinly gilt, is based on this reaction. The experiment is finished in less than a minute ; if a manganese salt is used the rings of manganese dioxide that are formed are more numerous than those of lead dioxide. If cobalt sulphate is used instead of manganese sulphate, rings of cobalt oxide do not form until after about twenty minutes.

The process of electroplating with silver, by the use of a solution of potassium-silver cyanide ($KAgCy_2$), which is such a common technical operation, is accomplished, according to Hittorf (*Über die Wanderungen der Ionen*, 2 Hälfte, p. 74 [*Ostwald's Klassiker*]), by the following reactions: the kation potassium (K) travels to the kathode, and the anion $AgCy_2$ to the anode; and each K ion precipitates silver on the kathode in accordance with the equation

$$K + KAgCy_2 = 2KCy + Ag,$$

while the anion $AgCy_2$ dissolves an atom of silver from the silver anode, and these re-form the complex cyanide ($AgKCy_2$) by combining with the $2KCy$ [produced in the way shown by the above equation]. If the anode consists of platinum, cyanogen gas is set free thereat from the anion $AgCy_2$, and the platinum becomes covered with silver cyanide [$AgCy$] which soon stops the current. Hittorf attributes the separation of the silver in a coherent and homogeneous form to the fact that the precipitation on the kathode is the result of a secondary reaction; it is this circumstance which gives its importance to the technical applications of the double cyanide as an electrolyte. For the silver which is separated from a solution of silver nitrate, as the result of the primary reaction of the current, forms branching crystals, which increase downwards, and are easily rubbed off the electrode.

The double potassium-gold cyanide, $KAuCy_2$, is employed in electroplating with gold, as the double silver cyanide is employed in plating with silver. The potassium cyanide process, which is the most recent method, for extracting gold depends on the electrolysis of this gold salt. This process has become so important that the value of the gold extracted by its use in 1894 in the Transvaal amounted to about £7,000,000, which is an output that places that republic in the forefront of the gold-producing countries at the present time. The gold, which is found in a very finely divided condition in the rocks, is dissolved by a dilute solution of potassium cyanide, in the presence of oxidising bodies, in accordance with the equation

$$4KCy + 2Au + H_2O + O = 2KAuCy_2 + 2KOH;$$

and the solution is subjected to electrolysis between steel anodes and lead kathodes. Prussian blue, which is again worked up into potassium cyanide, is formed on the anodes; the leaden kathodes, encrusted with gold, are taken to the furnace-hearths, whereon the gold remains after the removal of the lead by cupellation.

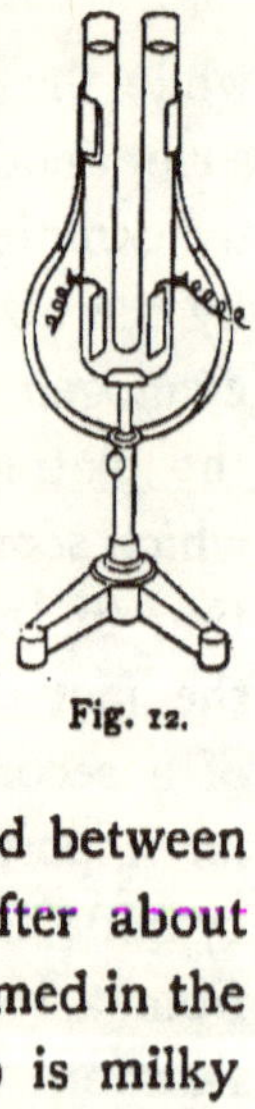

The electrolysis of a solution of potassium ferrocyanide, K_4FeCy_6, containing 10 c.c. of a saturated solution of the salt mixed with 200 c.c. of water, and acidified by hydrocyanic acid, shows how complicated the secondary reactions in an electrolytic process may become. The current from five accumulators is passed through the solution placed between platinum electrodes in a U-tube (*see* fig. 12). After about twenty minutes, Prussian blue, $(Fe_2)_2(FeCy_6)_3$, has formed in the anode limb, while the liquid in the kathode limb is milky from small uprising bubbles of hydrogen. According to Hittorf (*loc. cit.*, p. 72), 4K goes to the kathode, where it reacts with water as shown by the equation

$$4K + 4H_2O = 4KOH + 4H,$$

and the ion $FeCy_6$ goes to the anode. If there is a sufficient supply of K_4FeCy_6, the anion reacts with that compound to

Fig. 12.

produce potassium ferricyanide, K_3FeCy_6, according to the equation

$$3K_4FeCy_6 + FeCy_6 = 4K_3FeCy_6.$$

But if the solution is as dilute as that mentioned above, the following processes occur at the anode:—

$$FeCy_6 + 2H_2O = H_4FeCy_6 + 2O,$$
and $$7H_4FeCy_6 + 2O = 24HCy + (Fe_2)(FeCy_6)_3 + 2H_2O.$$

The electrolysis of a concentrated solution of sodium acetate, $CH_3.COONa$, produces a combustible gas at each

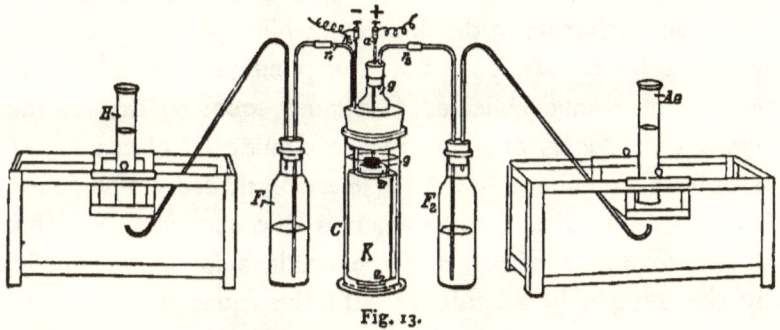

Fig. 13.

electrode. A cylinder of about one litre capacity, such as was used by v. Hofmann in his lectures, serves as the decomposition-cell (see C, fig. 13). The porous cell z is placed in this cylinder, and over the upper edge of the cylinder is inverted the bell-shaped vessel g (made by cutting off the bottom of a flask). The cylindrical piece of copper foil K, which is soldered to the conducting strip k, is the kathode; and the platinum plate fastened to the wire a is the anode. The exit tube for gas r_1 is connected with a flask, F_1, containing water; and the exit tube r_2 is connected with another flask, F_2, of the same size, but containing a volume of baryta water equal to the volume of water in F_1. All the corks must, of course, fit very tightly. The current from five accumulators is sufficient for the experiment. Hydrogen, produced in the secondary reaction

$$2Na + 2H_2O = 2NaOH + 2H,$$

escapes by r_1. According to Jahn (*Grundriss der Elektrochemie*, p. 292 [1895]), the following processes take place at the anode:

$$2CH_3.COO + H_2O = 2CH_3.COOH + O,$$
and $$2CH_3.COOH + O = C_2H_6 + 2CO_2 + H_2O.$$

The gases ethane (C_2H_6) and carbon dioxide (CO_2) escape by r_2; the latter gas is absorbed by the baryta water (the absorption is shown by the formation of a white precipitate of barium carbonate in the flask F_2), while the ethane collects in the cylinder, Ae, placed in the pneumatic trough. The volume of ethane collected is almost equal to that of the hydrogen which is obtained in the cylinder H placed in the trough on the other side of the electrolytic cell. The deficit in the volume of ethane obtained is due, according to Jahn, to the oxidation of some of the acetic acid, at the anode, by the oxygen in accordance with the equation

$$CH_3.COOH + 4O = 2CO_2 + 2H_2O.$$

The hydrogen and ethane may be distinguished by the luminosities of their flames; or, better, by the fact that the ethane takes fire quietly, while the hydrogen ignites with slight explosion. The experiment is of especial interest in organic chemistry.

It is to be remarked, in concluding this chapter, that endeavours are being made to turn to account the secondary reactions of electrolytic processes between indifferent electrodes for the practical preparation of organic compounds. If a solution of potassium sulphocyanide (1 : 5) is electrolysed, in the apparatus represented in figure 8 (p. 17), by the current from twelve accumulators, hydrogen is given off at the kathode, while the oxygen at the anode oxidises the sulphocyanic acid to *kanarin*, which is a yellow colouring matter used in dyeing, probably in accordance with the equation

$$6HCNS + 11O + H_2O = C_6H_4O_2N_4S_5 + H_2SO_4 + 2HNO_3.$$

The kanarin soon separates in yellow flocks; and after fifteen minutes enough will be obtained for a dyeing experiment, for which purpose the substance is dissolved in an alkali solution.

Every one knows that iodoform is produced by the reaction of iodine with a solution of sodium carbonate containing alcohol, at 60° to 80°. The same compound is formed electrolytically by separating iodine, at the anode, from potassium iodide, under conditions such that the occurrence of a secondary reaction with alcohol in presence of sodium carbonate is possible. The apparatus represented in figure 3 (p. 9) may be used as the electrolytic cell. The tube is filled with a solution of 5 grams potassium iodide and 10 grams sodium carbonate in 100 c.c. of water to which 10 c.c. of alcohol have been added; the apparatus is immersed in warm water in a beaker, and the current from a battery of four accumulators is sent through the liquid. After five minutes the smell of iodoform can be perceived in the limb which contains the anode; and after twenty minutes the bulb of the U-tube is partly filled with powdery iodoform.

The electrical method has found an extensive application in the manufacture of those organic compounds which are required as intermediary products in the making of aniline colours. The compounds wherewith the processes of change commence have to be reduced at one time and oxidised at another; they are, therefore, placed either near the kathode, or near the anode, of an electrolytic cell, in which an acid or an alkali, according to the special requirements of the case, is subjected to electrolysis. The intensity of the current, the concentration, and the temperature must be regulated in a special way for each preparation. Nitro-compounds, for instance, are reduced to hydrazo-compounds in alkaline solutions, and to amido-compounds in acid solutions.

An example of the oxidising effects of the electric current is furnished by the conversion of aniline into aniline black, in accordance with the equation

$$2C_6H_5NH_2 + O_2 = 2H_2O + C_{12}H_{10}N_2.$$

The experiment may be performed by mixing 100 grams of aniline with 100 c.c. of pure hydrochloric acid (spec. grav. 1·119), dissolving 50 grams of the aniline hydrochloride which crystallises out on cooling in 500 c.c. of water, adding 20 grams potassium chlorate, placing this mixture in an electrolytic cell like that used in the experiment illustrative of electrical bleaching (p. 12), with a piece of white woollen cloth, which has been boiled in dilute soda solution, stretched on the anode-plate, and sending the current from four accumulators through the liquid; after about thirty minutes those parts of the woollen cloth which are in direct contact with the electrode are coloured greenish black. The electrolytic formation of aniline black may be demonstrated more simply as follows. A mixture of 19 grams aniline and 22 grams toluidine is neutralised by 24 grams glacial acetic acid; the acetate is dissolved in 200 c.c. water, and a few c.c. of this solution are placed in the anode-limb of the U-tube shown in fig. 12 (p. 27), which is filled to above the electrodes with a concentrated solution of common salt. When the current from six accumulators has passed for a few minutes the contents of the anode-limb become deep black, because of the reaction brought about by the chlorine which is liberated.

CHAPTER II.

FARADAY'S LAW.

THE numerous experiments that have been performed so far have made evident how varying the action of an electrolysing current may be. Once more it must be emphasised that the primary action of the current always is to separate the electrolyte into two parts, one of which parts, that, namely, which is of a metallic character, travels towards the kathode, while the other, which moves towards the anode, consists of the rest of the compound.

It occurred to Faraday, and the idea was a happy one, to send the same current through a series of electrolytic cells arranged one behind the other and containing different electrolytes. It thus became possible to make a quantitative comparison of the changes brought about by the same quantity of electricity in motion. The result of the experiments found expression, in 1833, in *Faraday's law of constant electrolytic action.* In the form given to it by H. von Helmholtz, the law states that *the same quantity of electricity* * *passing through an electrolyte either sets free, or transfers to other combinations, always the same number of valencies.*

The following experimental arrangement is suitable for deducing the law for the kations. In the circuit of the current from five accumulators are included a rheostat, by

---

* Not to be confounded with the energy of the current.

means of which the current is at first diminished, a Hofmann's apparatus for the electrolysis of water, and four glass troughs: the Hofmann's apparatus and two of the troughs are represented in figure 14. The electrode-plates a and k project downwards in each trough; the conducting wires of these plates are connected by binding screws with the slips of copper $S\,S$. The kathodes are composed wholly of platinum; they must be cleaned carefully, by fuming nitric acid, and weighed, accurately to a centigram, before the experiment begins. Either platinum, or the metal which is contained in the electrolyte, may be used for making the anode in each cell. In selecting electrolytes, care must be taken that the metals deposited on the kathodes, shall adhere firmly to the platinum plates, that these deposited metals shall not suffer oxidation, at any rate not during the time required for weighing, and that the valencies of the atoms of the metals shall be as different as possible. The following electrolytes are to be recommended:—

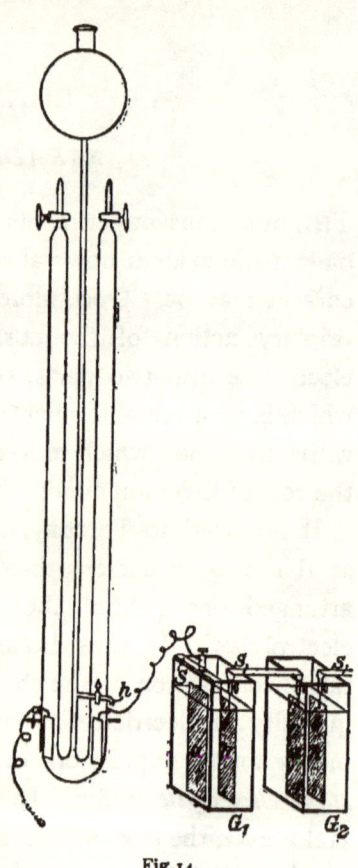

Fig. 14.

(1) A solution of potassium-silver cyanide, obtained by dissolving 3 grams silver nitrate and 5 grams potassium cyanide in 200 c.c. water.

(2) A solution of cuprous chloride, prepared by dissolving

3 grams of the salt sold by the chemical dealers, after washing on a filter with v ater, in hydrochloric acid and diluting to 200 c.c.

(3) A solution of copper sulphate, made by adding 100 c.c. water and 15 c.c. nitric acid to 100 c.c. of a saturated solution.

(4) A solution of tetrachloride of tin, prepared by dissolving 1 gram tin-foil in hydrochloric acid, adding a few drops of bromine, evaporating until almost every trace of free acid is removed, and then adding 100 c.c. water and 100 c.c. of a saturated solution of ammonium oxalate.

The electrolysis is allowed to proceed for about thirty minutes; the kathode-plates are then rinsed with water, dried thoroughly with alcohol and with ether, and weighed. In *Table I.* are given the results of a carefully conducted experiment.

TABLE I.

ELECTROLYTE.	I. Diluted sulphuric acid (1 : 12).	II. $KAgCy_2$.	III. CuCl	IV. $CuSO_4$	V. $SnCl_4$
Material which formed the electrodes . .	− + Pt Pt	− + Pt Ag	− + Pt Cu	− + Pt Cu	− + Pt Pt
Quantity of kation separated . .	67 c.c. = 6·002 mgm. H	650 mgm. Ag	380 mgm. Cu	190 mgm. Cu	170 mgm. Sn
Weight of kation referred to 1 mgm. H.	1 mgm. H	108·2 mgm. Ag	63·6 mgm. Cu	31·8 mgm. Cu	28·3 mgm. Sn
Atomic weight . .	1	107·6	63·3	63·3	117·8
Percentage error . .	—	+ 0·6	+ 0·4	+ 0·4	− 4

When the numbers which express the weights of the kations that have been separated are referred to a unit weight of hydrogen, they are found to be very nearly the same as the quantities obtained by dividing each atomic weight by the valency of the atom; for in columns II. and III. the silver and copper atoms are monovalent, in column IV. the copper atoms

CHAP. II.] FARADAY'S LAW. 35

are divalent, and in column V. the atoms of tin are tetravalent. The law of Faraday is demonstrated satisfactorily by the experiments. Moreover, the results show that the quantities of the kations that have been separated are proportional to the quantity of the current, and also to the duration of the action of the current.

The more accurate experiments of F. and W. Kohlrausch have shown that 0·3281 mgm. of copper was separated from cupric salts by one *coulomb** of electricity; or, that 31·65 grams of copper were separated by 96,465 (in round numbers 96,500) coulombs. This quantity of electricity expresses the *electro-chemical equivalent*, that is to say, the number of coulombs which causes the separation, in one second, of that fraction of the atomic weight of a metal, or of the [formula-] weight of an anion-group, expressed in grams, which corresponds with a single valency.†

The measurement of the intensity of a current [that is, the quantity of electricity which passes through any cross-section of the conductor in one second] by means of voltameters depends upon the exact proportionality between the quantity of electricity and the quantity of the ions that are separated. One coulomb separates 1·1181 mgm. silver, per second, in a

---

* If the unit quantity of electricity, one *coulomb*, flows through any given section of the conductor in one second, the intensity of the current is one *ampère*. The unit of potential, the *volt*, is so fixed that one coulomb of electricity is driven by it, per second, through a resistance of one *ohm*, that is, through a column of mercury 106·3 centims. long and 1 sq. mm. section (at 0°); the coulomb is so adjusted that 1 coulomb × 1 volt, that is, the unit of electrical energy, is equal to the work of 10^7 *ergs*. This quantity of energy is called a *watt*. Joule

† The statement in the text may be put thus. The electro-chemical equivalent is the number of coulombs that causes the separation of a *gram-equivalent* of a metal, or anion-group, in one second; a gram-equivalent being the quotient obtained by dividing the atomic weight of the metal, or the formula-weight of the anion-group, taken in grams, by the valency of the metal or anion-group.—[TR.]

silver-voltameter, and sets free 0·174 c.c. explosive gas [hydrogen and oxygen in the ratio of 2 to 1 by volume] (reduced to 0° and 760 mm. pressure) in the same time in a water-voltameter. Supposing that 20 c.c. hydrogen and 10 c.c. oxygen (referred to normal temperature and pressure) were obtained, in an apparatus for electrolysing water, in three minutes, this is equal to a production of 0·1667 c.c. explosive gas per second ; and, therefore, the quantity of electricity was 0·1667 / 0·174, that is 0·958 coulomb.

If one has three apparatuses of nearly equal size for the decomposition of water, whose resistances are equal (say 10 ohms each), then the law of the division of the current may be experimentally deduced in a few minutes by using the arrangement shown in figure 15 (see Grätz, *Die Elektricität und ihre Anwendungen*, p. 127 [1895]). *B* is a battery of twelve accumulators ; the resistance w_2 in the current-path $a\,A_2\,w_2\,b$ amounts to 100 ohms, and the resistance w_3 in the current-path $a\,A_3\,w_3\,b$ to 460 ohms. In an experiment there were obtained 42·5 c.c. explosive gas [mixture of two volumes hydrogen with one volume oxygen] in A_1, 34·5 c.c. in A_2, and 7·5 c.c. in A_3. Hence it follows that

$$i_1 = i_2 + i_3,$$

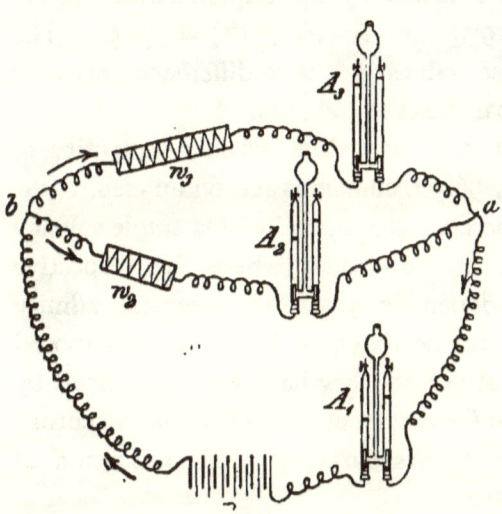

Fig. 15.

FARADAY'S LAW.

where the quantity of the current passing by the path $b\,B\,A_1\,a$ is represented by i_1, the quantity passing along the path $a\,A_2\,w_2\,b$ is represented by i_2, and i_3 is the quantity passing by the path $a\,A_3\,w_3\,b$. If the resistance of one of the decomposition-apparatuses is expressed by W (the three have all the same resistance), then the equation ought to hold good

$$i_2(w_2 + W) = i_3(w_3 + W).$$

The equation is fairly satisfied by the experimental results; for $i_2\,(w_2 + W) = 379 \cdot 5$, and $i_3\,(w_3 + W) = 352 \cdot 5$. The difference between these values is due to differences between the decomposition-apparatuses A_1, A_2, and A_3.

It follows from Faraday's law that when a quantity of electricity measured by 96,500 coulombs acts on an electrolyte, such quantities of the ions as correspond with a single valency always travel towards the electrodes where the respective actions take place, independently of the chemical affinity whereby these ions are bound together in the undecomposed molecules. The accuracy of the law has been established by H. von Helmholtz even for a current so feeble that a century would be required for it to separate a single milligram of explosive gas from water.

In the *Faraday Lecture* delivered by him in London on April 5th, 1881 [see *Journal of the Chemical Society*, 39, p. 277], H. von Helmholtz laid the foundations of a new electrochemical theory which explains the facts embraced by Faraday's law.

The most important of these facts may be stated thus: *Every single valency of an elementary or compound ion is charged with exactly the same quantity of positive or negative electricity, which behaves as if it were an electrical atom that cannot be further divided.* When, therefore, the circuit is closed, the kations in the electrolytic cell, being charged positively, are attracted to the kathode, and the negatively charged anions are attracted to the anode. If it is then possible for the ions to be set free, this does not occur throughout the whole mass of the electrolyte, but always only at the electrodes, and it is brought about by those quantities of electricity of opposite kinds which are carried to the electrodes by the current neutralising one another.* The result is that the ions cease to be in the ionic condition. This happens in the separation of the heavy metals on the kathode. But if the anion causes solution of the material of the anode to take place, that is to say, if it brings that material into the ionic condition, then a quantity of the metal of the anode, which quantity depends on the valency of that anion, becomes charged positively at the expense of the current. For instance, if the SO_4 ion dissolves one atom of copper from a copper anode, two positive charges are used, and the copper atom becomes a copper ion. When, therefore, a solution of copper sulphate is electrolysed between copper electrodes, the current produces new kations by charging the atoms of the metallic copper at the anode, and an equal quantity of positive electricity is given up at the kathode whereby the kations are un-ionised and become metallic atoms. If the ions react with the water, then negative hydroxyl ions are formed at

* The words used by von Helmholtz are these: "The same definite quantity of either positive or negative electricity moves always with each univalent ion, or with every unit of affinity of a multivalent ion, and accompanies it during all its motions through the interior of the electrolytic fluid. This quantity we may call the electric charge of the atom. . . . If we accept the hypothesis that the elementary substances are composed of atoms, we cannot avoid concluding that electricity also, positive as well as negative, is divided into definite elementary portions, which behave like atoms of electricity. As long as it moves about in the electrolytic fluid, each ion remains united with its electric equivalent, or equivalents. At the surface of the electrodes decomposition can take place if there is a sufficient electromotive force, and then the ions give off their electric charges and become electrically neutral" (*loc. cit.*, pp. 289-90).—[TR.]

the kathode, and positive hydrogen ions are formed at the anode, from the molecules of water, at the cost of the electrolysing current. Consequently, when the anion SO_4 appears at a platinum anode, two positively charged atoms of hydrogen, and also an atom of oxygen, are set free from the molecule of water; and when a potassium ion, or a sodium ion, appears at the kathode, a single hydroxyl group becomes negatively charged at the kathode, and this group plays the part of the anion necessary to the alkali ion, while an atom of hydrogen is disengaged at the kathode.

H. von Helmholtz has thus made clear wherein consists the process of the passage of electricity through a conductor of the second class, which must always be a chemical compound. At the same time, by the hypothesis that equal quantities of electricity cling to ions of equal valency, he has explained why the weights of the substances that are concerned in those chemical changes which are brought about by equal quantities of current electricity are always in the ratio of the equivalent weights of these substances. Moreover, it is possible to understand how isomeric ions can differ qualitatively, for instance in their colours; why the ferro-ion should be green while the ferri-ion is yellowish red, or why the MnO_4 ion of permanganic acid ($HMnO_4$) is violet and the MnO_4 ion of manganic acid (H_2MnO_4) is green. The qualities of the ions are dependent on their energy-content, and this, in turn, is conditioned by the number of the valencies, and consequently also by the number of the electric charges. (For more details see Chapter V.)

The positive charge on a hydrogen ion can be calculated approximately by considering that 1 mgm. of hydrogen is separated by 96,465 coulombs, and by making the assumption, which is based on definite facts, that this quantity of hydrogen consists of 1.2×10^{21} atoms. Hence a hydrogen ion must be charged with $96,465. / (1\cdot2 \times 10^{21}) = 8 \times 10^{-20}$ coulombs $= 8 \times 10^{-21}$ absolute units; this quantity may be taken to represent the absolute charge of a single valency.

It is certainly still doubtful how the details of the process of the neutralisation of ions at the electrodes are to be pictured in the mind. Two opinions exist. Either the ion becomes actually deprived of electricity when it has given up its proper charge, or it is furnished with a charge of the opposite kind in consequence of the use at the electrode of the double quantity of charge, and then it unites with an unchanged ion to form a molecule consisting of two oppositely charged atoms (for instance, $\overset{+}{H}\ \overset{-}{H}$). The latter hypothesis is hard to reconcile with the generally acknowledged monatomicity of the metallic molecules, and it tends on the whole to the final identification of electrical and chemical energy, an identification which was made years ago by Berzelius. But as our knowledge of the nature of the two forms of energy is still very defective, it seems better to adopt the first hypothesis, which, moreover, is the simpler of the two, and has already been employed in the explanations that have been given.

CHAPTER III.

HITTORF'S TRANSPORT-NUMBERS.

WHEN a current, which is not very strong passes through a solution of copper sulphate, for a long time, between copper electrodes placed vertically, no further change seems to take place beyond the travelling of the copper with the positive current from the anode to the kathode; the anode loses as much copper as the kathode gains. Nevertheless, if the electrodes are connected with a galvanometer after the passage of the current, a polarisation-current is observed in the direction opposite to that of the primary current. Now the polarisation-current cannot be caused by gases, as is the case in the electrolysis of dilute sulphuric acid between platinum electrodes, for gases do not appear on these electrodes if the primary current is made sufficiently weak. The primary current must, therefore, have caused changes of some kind in the copper sulphate solution itself which give rise to the polarisation-current. It was soon discovered that the changes in the copper sulphate solution consisted in an increase of the concentration of the solution at the anode and a decrease at the kathode, while the total quantity of copper sulphate in the solution remained constant.

In the years 1853-59 Hittorf made a study of a great many cases of those changes which occur in the concentrations of electrolytes at the electrodes. Hittorf's papers *Über die*

Wanderungen der Ionen are collected in Nos. 21 and 22 of *Ostwald's Klassikern*. The results of his laborious investigations were not properly appreciated until recently.

The decomposition-cells used by Hittorf were constructed so that the electrodes lay horizontally over one another, and the solution could be removed after the electrolysis in layers, without the risk of getting mixed, and each of these layers could then be analysed separately.

These changes of concentration can easily be made visible by means of the apparatus shown in figure 16. A piece of glass tubing, 30 centims. long and 3 centims. diameter, is closed at both ends by corks through which pass the thick conducting wires a and k, and these wires are soldered to perforated, sieve-like plates of copper. If the poles of a battery of five accumulators are attached to the conducting wires, and a rheostat is included in the circuit by means of which the intensity of the current is so reduced that gas is not given off, the liquid which surrounds the kathode, k, becomes quite colourless after some minutes.

The changes of concentration calculated from the data in one of Hittorf's experiments are represented by the device in figure 16. To make the illustration clearer, the percentage numbers are expressed by the corresponding numbers of ions, the anions being represented by the white dots and the kations by the black dots. The horizontal line separates the layer round the kathode from that around the anode. The liquid is homogeneous before electrolysis begins, and we may suppose that 9 kations and 9 anions are present in each layer. When the current has been passing for a certain time, 6 copper atoms are separated on the kathode, and an equal number of copper ions is dissolved by the SO_4 ions. While, however, only 5 Cu and 5 SO_4 are found on the kathode side, 7 Cu and 7 SO_4, besides the 6 re-formed

$CuSO_4$, are found on the anode side. If there were only a wandering of the anions during electrolysis, there would be $9 + 6 = 15$ $CuSO_4$, in all, in the anode layer, and there would, therefore, be $9 - 6 = 3$ $CuSO_4$ in the kathode layer. If, on the other hand, only 6 Cu ions had travelled from the anode layer to the kathode layer, to the 6 SO_4 ions that had there become free, then there would have been 9 Cu and 9 SO_4 in each layer as there was before electrolysis began. As a matter of fact, 5 Cu and 5 SO_4 are found at the kathode, and $7 + 6$ Cu and $7 + 6$ SO_4 at the anode. Both kinds of ions have been transported, the Cu ions towards the kathode and the SO_4 ions towards the anode, and, moreover, 2 Cu have moved from beneath upwards and 4 SO_4 from above downwards. Six copper atoms being set free at the kathode, 2 Cu ions move upwards, while the 4 remaining free SO_4 ions move downwards, so that there are, in all, 6 free SO_4 ions under the line, and these are provided with Cu ions at the cost of the anode. It may be said

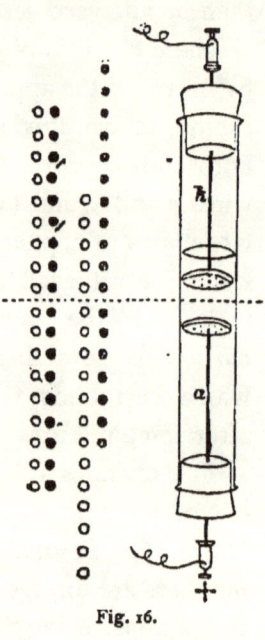

Fig. 16.

that a Cu ion will pass over two of six spaces, and a SO_4 ion will pass over four in the same time. The quotients $\frac{2}{6} = 0.33$ and $\frac{4}{6} = 0.66$ are called by Hittorf the *transport-numbers* (*die Überführungszahlen*) for the kation Cu and the anion SO_4, respectively. These numbers express the quantities of both ions transported for each copper atom that is separated; or they may be taken to represent the distance passed over by one ion, divided by the sum of the distances traversed by both ions.

If the transport-number of the anions is expressed by n, that of the kations is $1-n$. The ratio $(1-n) : n$ must evidently be the ratio of the velocities, u and v, of the kations and anions under the influence of the difference of potential at the electrodes. Consequently the equation holds good

$$\frac{1-n}{n} = \frac{u}{v}$$

Hittorf's researches have shown the ratio u/v to be independent of the differences of potential at the electrodes, and also, within certain limits, independent of the concentrations of the solutions. The effect of temperature has shown itself to be inappreciable, if the experiments are made at about ordinary temperatures. But at higher temperatures the difference between u and v decreases more and more.

CHAPTER IV.

THE LAW OF KOHLRAUSCH.

IN his communications on the migrations of the ions, Hittorf repeatedly drew attention to the possibility of obtaining fuller insight into the nature of electrolysis by making measurements of the *specific conductivities* (that is, the reciprocals of the resistances) of electrolytes. But it was a long time before a practicable method for measuring the resistances of electrolytic solutions could be found, because gases are generally produced by the passage of the current through these solutions, and those gases give rise to an opposing electromotive force the magnitude of which is subject to variations.

F. Kohlrausch discovered a method in 1880. The principle of his method is the same as that which is applied to the measurement of the resistances of wires by the use of Wheatstone's bridge. The influence of polarisation is eliminated by the employment of an alternating current from an induction-machine, and a telephone is used as indicator in place of a galvanometer. The solution under examination is placed in a cell between platinised platinum electrodes [that is, platinum plates covered with a fine coating of platinum black, by electrolysing a very dilute solution of platinum chloride between them].

The method of measurement is represented diagrammatically in figure 17. *G* is an accumulator which works the induction-machine *J*; *A B C D* represents the branching of the current, and *a*, *b*, and *c* are Siemens' rheostats. The resistances in *a* and *b* remain constant; they may bear the relation to one another of 1 to 100. *Z* represents the cell that is filled with the electrolyte whose resistance is to be measured. If the resistance in *c* is now altered until the telephone, which is included in the branch *C D*, sounds, then the resistance in *Z* must be 100 times that in *c*.

The resistance of an electrolytic liquid is generally expressed in mercury units, one of which is equal to 0·94 ohm: the specific resistance, *s*, is found by taking into consideration the dimensions of the cell; *s* expresses the number of resistance-units of a thread of the liquid 1 metre long and 1 sq. mm. cross-section compared with an equal thread of mercury.

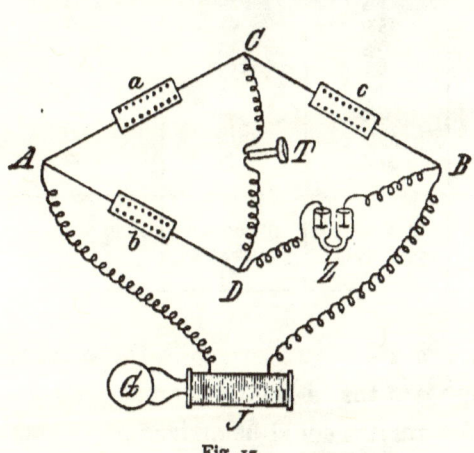

Fig. 17.

The *specific conductivity* of the liquid, $L = 1/s$, is found from *s*.

The conductivities of electrolytic solutions are much smaller than those of metals. The following table (*Table II.*) is drawn up from the results of Kohlrausch and Grotrian. The compound used as electrolyte is stated in column 1; the percentage contents of the solution (p) are given in column 2; column 3 gives the resistance (r) of one c.c. of the solution in ohms, at 18°; column 4 shows the conductivity at 18° (l) referred to that of mercury = 10,000,000; and in column 5 are given the increases of conductivity (Δl), per degree of temperature, in percentages of l.

TABLE II.

Electrolyte.	p.	r.	l.	Δl.
H_2SO_4	5	4·8	195	1·21
,,	30	1·4	691	1·62
HCl	5	2·5	369	1·59
,,	10	1·6	590	1·57
,,	20	1·3	713	1·55
,,	30	1·5	620	1·53
NaCl	5	15·0	63	2·20
,,	10	8·3	113	2·10
,,	15	6·1	153	2·10
,,	25	4·7	200	2·30
NaOH	17	2·9	326	—
$CuSO_4$	5	53·3	18	2·20
,,	10	31·4	30	2·20
Mercury	. . . .	0·0000943	10,000,000	
Copper	. . . .	0·0000017	550,000,000	

Now, just as in the case of Faraday's law, so in the present case, it was not to be supposed that a law to express the conductivities of different electrolytes could be arrived at as long as the concentrations of the solutions were stated merely in percentages; Kohlrausch, therefore, referred the values of L to equimolecular solutions, and so obtained the *molecular conductivities* λ. An example will make the conception of molecular conductivity more intelligible. The specific resistance, s, of a solution of potassium chloride containing 5 grams of the salt in 100 c.c. was found to be 160,256 in mercury units at 0°. Hence $L = 62·4 \times 10^{-7}$. As 5 grams of potassium chloride are contained in 100 c.c. of the solution, one gram-molecule, that is 74·5 grams [KCl = 74·5], would be contained in 1490 c.c. of an equally concentrated solution. It follows from the value found for L that the conductivity of one c.c. of the solution is $62\cdot 4 \times 10^{-3}$; hence

$$\lambda = 62\cdot 4 \times 1490 \times 10^{-3} = 93.$$

If, then, 1490 c.c. of the solution were contained in a cell between two electrodes of 1490 sq. centims. surface and placed one centim. apart, the resistance would be $\frac{1}{93}$ of a mercury unit.

The molecular conductivity of a solution of potassium chloride containing one gram-molecule of the salt in 1·49 litres is written in this form—

$$\lambda_{1\cdot 49} = 93.$$

If V is taken as a general symbol for the number of litres wherein one gram-molecule of the electrolyte is dissolved, then

$$\lambda_V = L.\ V.\ 10^7.$$

Kohlrausch's results give the following numbers for solutions of potassium chloride of different concentrations, at 18°:—

TABLE III.

74·5 gram KCl dissolved in	$s.$	$L.$	$\lambda.$
0·33 litre	0·00399 × 10⁷	250·0 × 10⁻⁷	82·7
1 ,,	0·01088 × 10⁷	91·9 × 10⁻⁷	91·9
2 litres	0·02087 × 10⁷	47·9 × 10⁻⁷	95·8
10 ,,	0·09360 × 10⁷	10·5 × 10⁻⁷	104·7
100 ,,	0·87184 × 10⁷	1·15 × 10⁻⁷	114·7
1000 ,,	8·38223 × 10⁷	0·119 × 10⁻⁷	119·8

These numbers show that the specific conductivity of an electrolyte decreases as concentration decreases, but that the conductivity does not decrease as rapidly as the concentration; for instance, the value for L_{10} is more than a tenth of that for L_1. If the tenth part of a thin column of a solution were made up to the original length of the column by the addition of pure water, the conductivity after dilution would be not exactly one-tenth, but more nearly 0·11, of the conductivity of the original column before dilution. Although only a tenth part of the original quantity of salt is present after dilution, nevertheless that tenth part of the salt is affected by the greater dilution in such a way that it conducts the current better than one would have expected; or, to use the language of the theory of Arrhenius, which will be considered in Chapter V., the relative number of *active molecules* has increased. The same thing finds expression in the values of λ given in the table. The following statement is deduced from these values :—

The molecular conductivity of an electrolyte increases with the dilution, and at a definite limit it reaches the maximum value λ_∞.

The maximum values can be calculated from the experimental data by the method given by Ostwald.* For potassium chloride the maximum value is 140.

By comparing the values of λ for very dilute solutions of two electrolytes with the same anion and different kations, on the one hand, and two electrolytes with another common anion and the same two kations as the previous pair, on the other hand [for instance, by comparing the values of λ for very dilute solutions of KCl and NaCl with the values

* The method is based on the fact that, so far as experiments have gone, every ion can be combined with another ion of such a nature that the maximum conductivity of the salt so produced is attained at a workable temperature. Sodium is the kation most generally employed, because solutions of all salts of sodium attain their maximum conductivities at dilutions within the limits of experimental error. And NO_3 is the most generally used anion, as the maximum conductivities of nitrates in solution are reached at workable dilutions. (*See* Ostwald's *Lehrbuch der allgemeinen Chemie*, vol. ii. (Part I.), pp. 673-4; also the article "Physical Methods" in Watts' *Dictionary of Chemistry*, new edition, vol. iv., p. 192.)—[TR.]

for very dilute solutions of KNO_3 and $NaNO_3$], Kohlrausch found that the change of kation was accompanied by an almost constant change in the value of λ independently of the nature of the anion. An example, considered below, will make the statement clearer. Kohlrausch concluded that *the maximum molecular conductivity*, λ_∞, *of an electrolyte is made up, additively, of two constants which can only represent the migration-velocities*, u *and* v, *of the ions.* On this hypothesis he set forth the equation

$$\lambda_\infty = u + v.$$

The values of u and v can be calculated easily; for $u : v = 1 - n : n$, where n is the transport-number of Hittorf for the anion (*see* p. 43). For potassium chloride, for example, at 25°, $\lambda_\infty = u + v = 140$; and, as $u : v = 0\cdot491 : 0\cdot509$, we get $u_K = 68\cdot6$, and $v_{Cl} = 71\cdot4$.

In Table IV., column I. gives the formulæ of certain electrolytes, column II. gives the values of Hittorf's transport-numbers for the kations of those electrolytes, column III. contains the most recently determined values of λ_∞ (Ostwald's *Lehrbuch der allgemeinen Chemie*, vol. ii., p. 675 of Part I.), and columns IV. and V. contain the values of u and v calculated in the way just described for the case of potassium chloride.

TABLE IV.

I. ELECTROLYTE.	II. $1 - n$.	III. λ_∞ at 25°.	IV. u at 25°.	V. v at 25°.
KCl	0·491	140·0	68·6	71·4
KNO_3	0·503	135·7	68·3	67·4
NaCl	0·380	120·0	45·6	74·4
$NaNO_3$	0·387	113·7	44·0	69·7
$AgClO_3$	0·499	117·2	58·5	58·7
$AgNO_3$	0·477	124·2	59·2	65·0

THE LAW OF KOHLRAUSCH.

These numbers show that $\lambda_{KCl} - \lambda_{NaCl} = 140 - 120 = 20$, and $\lambda_{KNO_3} - \lambda_{NaNO_3} = 135\cdot7 - 113\cdot7 = 22$. The differences are nearly the same. Now if the hypothesis of Kohlrausch is correct, if the formula $\lambda_\infty = u + v$ always holds good, the value of λ_∞ found empirically for one electrolyte must agree with the sum of the mean values for u and v, which are calculated from empirical data for n and λ_∞ for other electrolytes. The value of v_{NO_3} for the electrolyte KNO_3 is $67\cdot4$, and the value of u_{Ag} for $AgClO_3$ is $58\cdot5$; hence $u_{Ag} + v_{NO_3} = 125\cdot9$; the actual determination of λ gave $\lambda_{AgNO_3} = 124\cdot2$, which agrees well with the calculated value. The formula $\lambda_\infty = u + v$ is thus the expression of a law. The law is called *the law of the independent migration-velocities of the ions.*

The following mean values have been determined by Kohlrausch (*Wied. Annalen*, **50**, p. 385 [1893]); the temperature was 18°.

For K	$u =$	60	For Cl	$v =$	62
„ Na	„ =	40	„ NO$_3$	„ =	58
„ Ag	„ =	52	„ ClO$_3$	„ =	52
„ H	„ =	290	„ OH	„ =	165

Solutions of medium concentration of normal salts formed of two monovalent ions give results which are in close keeping with those demanded by the law of Kohlrausch, as do also certain strong mono-acid bases and monobasic acids. But the experimentally determined molecular conductivities of electrolytes with polyvalent ions are smaller, even when working with very dilute solutions, than the values required by the law; Ostwald has, therefore, given the following more general form to the law,

$$\lambda = a(u + v),$$

where a has the value of a proper fraction. The experimental data (which it must be confessed are very meagre as yet) always showed that the deviations became smaller as the dilution increased, and they indicated that a would become equal to unity at an infinite dilution. But the degree of dilution cannot pass beyond a certain limit if actual measurements are to be obtained. In such cases the value of u must be obtained by the aid of the value of λ_∞, as can be done accurately from the chloride or nitrate of the kation in question; and the value of v must be obtained from determinations of λ_∞ made by means of the sodium or potassium salt of the anion concerned.

The numbers given in *Table II.* (p. 47) show that λ must increase with rise of temperature. The following results were obtained for a $\frac{1}{30}$ normal solution of potassium chloride: $\lambda = 112\cdot2$ at $18°$, $\lambda = 129\cdot7$ at $25°$. Hence, by the equation established by Kohlrausch,

$$\lambda_{25°} = \lambda_{18°} [1 + \beta(t-18)],$$

the temperature-coefficient is $\beta = 0\cdot022282$. In most cases λ increases by about $2\cdot5$ per cent. for each $1°$ of temperature-increase, when working with very dilute solutions.

More details concerning the connections between the velocities of ions and chemical constitution will be found in Ostwald's *Lehrbuch der allgemeinen Chemie*, vol. ii., Part I. (1893).

There are two experiments which, although not quite free from objection, may be used to illustrate the law of Kohlrausch to a certain extent. A U-tube with platinum electrodes (*see* fig. 12, p. 27) is used as the decomposition-cell in the first experiment; and the current is passed through equimolecular solutions of two sodium salts, the velocities of the anions of which differ as much as possible, a moderately sensitive galvanometer with a vertical needle being intercalated. The solutions used are one of sodium acetate $84 : 100$ ($v_{C_2H_3O_2} = 38\cdot4$), and one of common salt $36 : 100$ ($v_{Cl} = 62$). The current is passed first through the sodium acetate solution, and then through the solution of common salt, the same U-tube being used. The needle shows a deviation in the latter case about three times greater than in the former, and this is due essentially to the greater velocity of migration of the chlorine ion in comparison with that of the anion $C_2H_3O_2$.

The second experiment furnishes an objective representation of the differences of ionic velocities. Figure 18 represents the principle of the arrangement devised by Lodge (*British Association Reports*, 1887, p. 389) for the direct ascertainment of

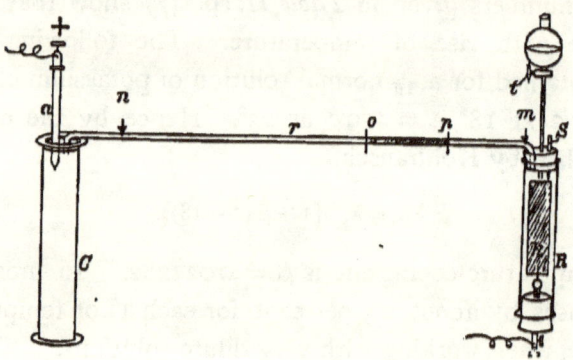

Fig. 18.

the velocities of the migration of ions; the apparatus figured here is copied, with some changes, from that used by him. A centimetre scale is cut, by a diamond, on a piece of glass tubing, r, 40 centims. long and 8 mm. wide, and the tube is bent, at right angles, at $1\frac{1}{2}$ centims. from either end. Ten grams of pure gelatin are heated with 140 c.c. of water on a water bath until the gelatin is dissolved; 7 grams of pure sodium chloride, and a few drops of a slightly alkaline solution of phenolphthaleïn, are then added, and the whole is thoroughly mixed and filtered through paper, and the rose-pink filtrate is poured into the tube r, where it soon solidifies. One end of the tube r is passed through a cork, which also carries a stoppered funnel, t, and a little piece of glass rod, s; this cork fits into the upper end of the tube R, the lower end of which is also closed by a cork that carries the platinum plate k, which serves as the kathode. By means of the stoppered funnel t, and the glass rod s, it is not difficult to fill the tube R with a solution of cupric chloride (1 : 10) so completely that all the air is driven out of this tube; if any air were left in R, some of the gelatin would be forced out of r when the current began to pass. The other end of the tube r dips into a glass cylinder, C, which is filled with diluted hydrochloric acid, wherein is placed a rod of gas carbon, a, that serves as the anode. The object of the whole arrangement is to show that, when the current passes, the hydrogen forces its way from C, and the chlorine from R, into the tube r, and that this is rendered visible by the decolourisation of the gelatin. It is necessary, however, to leave the apparatus at rest for at least twenty-five hours before the battery is connected; for the liquids in R and C gradually diffuse into the tube r, with the result that the gelatin is decolourised at the end near C by the neutralisation of the alkali by the hydrochloric acid, and that the pink colour is replaced in the part of the tube near R by a slightly blue turbidity due to copper hydroxide, which is formed by the reaction between the copper chloride that diffuses into the gelatin and the caustic soda that it meets there ($CuCl_2 + 2NaOH = CuO_2H_2 + 2NaCl$).

The following numbers show the distances passed over in stated times by the solutions of hydrochloric acid and cupric chloride respectively, the temperature being 20°:—

	1 hour.	4 hours.	25 hours.	36 hours.
Distance by which the hydrochloric acid advanced.	1 centim.	2 centims.	5 centims.	6 centims.
Distance by which the cupric chloride solution advanced.	$\frac{1}{2}$,,	1 ,,	$2\frac{1}{2}$,,	3 ,,

THE LAW OF KOHLRAUSCH.

These numbers agree with the law of Stefan: $h = a \sqrt{t}$, where h is the distance passed over by the diffusing liquid in the number of hours t, and a is a constant. For dilute hydrochloric acid $a = 1$, and for the cupric chloride solution $a = \frac{1}{2}$.

After twenty-five hours, and when the gelatin in the part of the tube mn is still rose-pink, a battery of ten accumulators is connected with the electrodes. The process of decolourisation now proceeds unequally. While the chlorine ions and the copper ions are un-ionised at a and k, and the current is thus allowed to pass through the apparatus, the hydrogen ions of the hydrochloric acid travel from a towards k, and the chlorine ions of the cupric chloride travel from k towards a; in two hours the hydrogen ions move over a space of 3 centims., and the chlorine ions over a space of 0·5 centim. The movements of these ions are attended by the decolourisation of the gelatin. After ten hours only the little part op remains red; the decolourisation has spread in this time to a distance of 18·8 centims. from the anode, and to 3·7 centims. from the kathode (the gelatin in the space mp is no longer blueish and turbid, but it is as colourless as in no). The red zone at last quite disappears, between the marks 31 and 32.

The decolourisation which is brought about by the current is due to the combination of the hydrogen ions, that come from the anode, with the hydroxyl of the base that is present in the gelatin, whereby water is formed, while the sodium atoms that remain over combine with the chlorine ions that approach from the kathode, and so produce a neutral salt. It thus comes about that, as the ions travel through the gelatin, the alkali is removed at both sides, and the phenolphthaleïn becomes colourless.

The decolourisation of the gelatin would have been accomplished by the diffusion process alone, in the time during which the current has been allowed to pass, to the distance of 0·9 centim. from the anode and 0·4 centim. from the kathode; these numbers must therefore be deducted from 18·8 and 3·7 respectively; the differences, 17·9 and 3·3 centims., express the ratio of the velocities of migration of the ions H and Cl under the influence of the current.

The experiment shows that the hydrogen ions advance towards the kathode about five times more quickly than the chlorine ions move towards the anode. Graham's experiments proved that a salt diffuses through a jelly at a rate scarcely less than that at which it diffuses through pure water; hence we may conclude that the numerical results of the present experiment would not be altered by substituting water for the gelatin.

By measuring the difference of potential at the ends of the tube r, it would be possible to express the migration-velocities of both ions in absolute units, that is, in centimetre/seconds per volt/seconds. According to Budde and Kohlrausch, the absolute values, U and V, are obtained by multiplying the relative quantities u and v by the factor 110×10^{-7}; these values are thus obtained:—

$$U_H = 0\cdot00352 \text{ and } V_{Cl} = 0\cdot00069 \text{ centim. at } 18°.$$

That is to say, the ions travel 0·00352 and 0·00069 centim., respectively, in one second, provided the difference of potential at the ends of the tube 40 centims. long is exactly 40 volts.

The following values are given by Kohlrausch (*Wied. Annal.*, **50**, p. 403) for very dilute solutions:—

$U_K = 0\cdot00066$ centim. $V_{NO_3} = 0\cdot00063$ centim.
$U_{Na} = 0\cdot00045$,, $V_{OH} = 0\cdot00181$,,
$U_{Ag} = 0\cdot00057$,,

The migration-velocities of polyvalent ions have not been determined as yet with sufficient accuracy.

CHAPTER V.

THE DISSOCIATION THEORY OF ARRHENIUS.

IN 1887 Svante Arrhenius put forward his *theory of the electrolytic dissociation of ions*, and thereby gave an explanation which might be accepted of those phenomena of electrolytic conductivity which had been summed up by Ostwald in the general formula $\lambda = a\,(u+v)$.

As the law of Kohlrausch tells that, under the influence of the electric current, the ions travel with definite velocities dependent on their chemical natures, and as it must be a matter of indifference which anion belongs to a determinate kation, Arrhenius asserted that *the molecules of electrolytes in aqueous solutions are already dissociated into their two ions which are loaded with their respective electric charges*; that electrolysis does not, therefore, require the previous splitting of the molecule by the electric current. While under ordinary conditions the ions move irregularly to and fro among the water molecules, and one ion will sometimes approach and sometimes recede from an ion of the opposite kind, if a difference of potential is established between the electrodes immersed in the solution, then, according to Arrhenius, the ions follow definite paths, the kation goes towards the kathode and the anion towards the anode, and they also move more quickly. The first work which the electrolysing current has to do is to overcome the frictional resistances experienced by the ions from the water molecules which stand in their way. These resistances vary in accordance with the nature of the ions, and the greater they are the smaller is the mobility of the ions, and therefore the smaller their migration-velocities. These electrolytic frictions are very considerable, according to Kohlrausch (*Wied. Annal.*, **50**, p. 407). To drive forward a gram-ion, in dilute solution, with a velocity of 1 centim. per second, requires a force equal to the weight of 984,000 / (E . U) or 984,000 / (E . V) kilograms, where E is the equivalent weight of the ion, and U, or V, is the velocity of migration of the kation, or anion, in absolute units (see end of last chapter). For 39·1 grams potassium, in dilute solution, would be required a weight of 984,000 / (39·1 × ·00066) = 38 × 10^6 kilograms. This very considerable quantity of work, which is used for the transport of the ions, and which forms a substantial part of the energy that is brought into the solution by the current, is transformed into heat; just as a portion of the current-energy, depending on the specific resistance of the conductor, is changed into heat in a metallic conductor.

But if the current is to pass continuously through an electrolyte it must perform a second portion of work at the electrodes; it must neutralise the ions that are attracted to the electrodes, either by drawing away the charges that cling to them with a definite intensity, or by forming new ions from the material of the electrodes or from the water, new anions for the kations and new kations for the anions.

Accordingly, the passage of a current through an electrolytic liquid is dependent on the presence of free ions in the liquid, and any undissociated [non-ionised] *molecules that may be in the liquid take no part in the carrying of the current.*

In the formula $\lambda = a(u+v)$, the factor a expresses what fraction of the theoretical value $\lambda\infty$ the experimentally determined value λ actually is. But in the dissociation theory

CHAP. V.] THE DISSOCIATION THEORY OF ARRHENIUS. 59

a acquires a more definite meaning. As the passage of the current is due altogether to the free ions, a expresses the fraction of the total number of molecules of the electrolyte which has undergone dissociation; a is, therefore, called *the dissociation-coefficient* [or *ionisation-coefficient*].* If, for instance, 100 gram-molecules of the electrolyte are dissolved in one litre of water, and 80 gram-molecules are dissociated, then $a = 0.8$. The remarkable fact that the value of λ increases as dilution increases, that is to say, the conductivity of the same quantity by weight of the electrolyte increases, is explained by Arrhenius by supposing that the continual addition of the solvent brings about the dissociation [or ionisation] of more molecules of the electrolyte, and therefore the production of more ions available for the carrying of the electricity. Arrhenius has expressed this view by saying that the number of *active molecules* is increased. If all the molecules are dissociated [or ionised] at a definite dilution, then the conductivity reaches its maximum at that dilution, and is expressed by the symbol λ_∞. In this case $a = 1$.

From the equations $\lambda = a(u + v)$ and $\lambda_\infty = u + v$, it follows that

$$a = \lambda/\lambda_\infty.$$

Such chemically pure liquids as condensed hydrogen chloride or 100 per cent. sulphuric acid do not conduct, because, according to the dissociation theory, their molecules are not ionised. For the same reason chemically pure water is not an electrolyte; for the water purified by Kohlrausch and Heydweiller, and distilled in a vacuum, showed a specific resistance at 18° of

* It seems distinctly better to use the terms *ionisation*, and *ionisation-coefficient*, rather than dissociation, and dissociation-coefficient. Dissociation has always been thought of as a process of separation of molecules into atoms, or atomic groups, such that the products are unconnected by any chemical bonds and can be separated from one another by diffusion. —[TR.]

24.75×10^{10} mercury units (*Sitzungsberichte der K. preuss. Akad. physik.-math. Klasse*, 1894, p. 295). A column of that water 1 mm. high would oppose a somewhat greater resistance to the passage of a current than that of a copper wire of equal thickness stretched three hundred times round the equator. The molecular conductivity of a litre of that water would be 0.404×10^{-4}; and, as $u_H = 290$ and $v_{OH} = 165$, the dissociation of the water would be so small that 1 gram H ions and 17 grams OH ions would be contained in $12\frac{1}{2}$ million litres.

Water as pure as this may, therefore, be regarded as a non-conductor; and it may be taken for granted that the water in a solution does not take a primary part in the electrolysis of the solution.

The circumstance that the most carefully distilled water is not a perfect non-conductor is due, according to the accurate observations of Warburg (*Wied. Annal.*, 1895, p. 396), to the presence of very minute quantities of impurities which are electrolytes, and which are extremely difficult to remove. The very small conductivities which are always found to be exhibited by such organic compounds as aniline, xylene, and turpentine-oil, however carefully the compounds may be purified, are to be assigned to a similar cause. Warburg recommends the removal of impurities of this kind by electrolysis, a process whereby the compounds are *purified electrically*.

In face of the facts that neither any simple electrolyte, nor pure water, conducts the current, it is extremely remarkable that aqueous solutions of these electrolytes should allow the current to pass. The water must be looked on as able to separate the molecule of an electrolyte into its two ions, and therefore to overcome the forces by which the ions are held together to form a molecule which appears neutral when viewed from without. As the occurrence of the purely physical process of the dissolution of an electrolyte is generally accompanied by the disappearance of heat, it is probable that this locking up of energy is connected with the work of dissociation. For instance, 8500 gram-calories, corresponding with 3600 kilogram-metres of work, are locked up during the dissolution of 1 gram-molecule (101 grams) of potassium nitrate. We have not as yet been able to penetrate further into the details of the mechanism of dissociation [that is, of the ionisation of electrolytes in aqueous solutions].

There are certain other liquids, besides water, which are able to cause the dissociation of electrolytes dissolved in them ; these liquids are compounds, like alcohol, which contain hydroxyl groups. Water, however, greatly surpasses all these liquids in its power of bringing about dissociation, and it is because of this, as well as because of its other exceptional properties, that water plays so important a part in the economy of nature.

While a solution of hydrogen chloride in water is an excellent conductor, the passage of a current, even a current of considerable intensity, is completely stopped by a solution of the same compound in chloroform. For, if a solution of dry hydrogen chloride in chloroform is placed in a U-tube (*see* fig. 12, p. 27) fitted with platinum electrodes, which are connected with a battery of ten accumulators, the needle of a galvanometer included in the circuit will not show the slightest deviation. The absence of ions in this solution is also the reason why litmus paper is unchanged when it is immersed in the liquid ; it is only after water has been added that the litmus is reddened.

There are certain aniline colouring-matters which behave towards water and towards other solvents similarly to hydrogen chloride. A very little [of the potassium salt of] *eosin*, $(C_{20}H_6Br_4O_5)K_2$, is shaken with a mixture of 20 c.c. ether and 1 c.c. alcohol, and the liquid is poured through a filter of the best Swedish paper. The filtrate is perfectly colourless, although it contains traces of a salt of potassium in solution. If a couple of c.c. of water are now added, and the liquid is shaken, the watery solution that settles to the bottom of the vessel looks rose-red by transmitted light, and shows a beautiful green fluorescence in reflected light. This production of colour is obviously to be referred to the dissociation of the saline compound of eosin by the water. The research of E. Buckingham (*Zeitschrift für physikal. Chemie*, 14, pp. 129-48 [1894]) shows that the fluorescence is certainly due to the complex anion of the eosin salt, for the fluorescence is the stronger the more the dissociation is promoted. *Methylene blue* ($[C_{16}H_{18}N_3S]Cl$), which is the chloride of a complicated kation, may be used to demonstrate, by a change of colour, the dissociating action of water. The compound dissolves very slightly in the mixture of alcohol and ether mentioned above, without colouring the solvent; but as soon as the solution is shaken with water, the compound is dissociated, and the free kations ($C_{16}H_{18}N_3S$) which are now present give to the water a deep blue colour, as intense as that of a saturated solution of copper sulphate. Violuric acid is also a suitable substance for recognising the dissociating action of water. This acid has the constitution

$$CO\begin{cases} NH.CO \\ \dot{C}=N.OH \\ NH.\dot{C}O. \end{cases}$$

It is prepared by heating a solution, as concentrated as possible, of 75 grams hydroxylamine hydrochloride with 145 grams alloxan for two to three hours, on a water-bath at 60° to 70°. An alcoholic solution of violuric acid is almost colourless; but when water is added, ionisation takes place, and the anion is coloured violet. A drop of hydrochloric acid suffices to reverse the dissociation, for the hydrochloric acid, being much more easily dissociated than the violuric acid, takes away the water from the latter.

The conductivity of electrolytes when fused can be understood, in so far as the heat used for melting also takes part in the work of dissociation.

There is, however, still a question to answer, and it is this: Whence are the electric charges of the ions, those constituents of the molecule of an electrolyte, derived? When the ions are not elementary atoms do they form themselves from such atoms, and must we think of these atoms as in themselves unelectrified?

These questions are not yet decided. The first step towards a solution has, however, been made by Ostwald (*Zeitschrift für physikal. Chemie*, 11, p. 501 [1893]), who ascertained the heats of ionisation of the elements, that is, the quantities of heat that become free, or are used, in the passage of gram-atoms of the elements into the ionic state.

A very brief account may be given here of the method by which Ostwald ascertained the value of j (the heat of ionisation) for copper.

When a current passes between copper electrodes through a solution of copper sulphate, copper is dissolved at the anode. The total quantity of heat, w, which is disengaged in this process is 10,200 gram-calories;* this is equal to the algebraic sum of two quantities, namely, the heat of ionisation, j, of copper, and that quantity of heat which is equivalent to the electrical energy, E, corresponding with the difference of potential, π, between the copper electrode and the (normal) solution of copper sulphate. Now as $\pi = +0·6$ volt,

---

* This is determined, indirectly, from the change that occurs in the potential-difference between the copper electrode and a solution of copper sulphate when the temperature changes.

according to measurements (carried out with a capillary electrometer) wherein the potential of the electrolyte = 0, and hence is smaller than the potential of the electrode, the quantity of energy set free, E, when one gram-atom of copper dissolves, is

$$E = 2 \times 96{,}500 \times 0{\cdot}6 \text{ volt-coulombs}$$
$$= 27{,}700 \text{ gram-calories.}$$

Hence, as the equation $w = E + j$ must hold good, it follows that

$$j = 10{,}200 - 27{,}700$$
$$= -17{,}500 \text{ gram-calories.}$$

This result means that 17,500 gram-calories disappear during the process of ionising one gram-atom of copper; so that the ion of copper is richer in energy, by this amount, than the gram-atom of copper.

It is very easy to deduce the value of j for zinc from the value for copper. The thermo-chemical equation Zn + CuSO Aq = ZnSO$_4$Aq + Cu + 50,100 calories tells that 50,100 calories are given out when one gram-ion of copper is un-ionised and at the same time one gram-atom of zinc is ionised. Now as 17,500 calories must be set free by the passage of one Cu ion into the non-ionised state, the process of ionising one gram-atom of zinc must produce 50,100 − 17,500 = 32,600 calories. Hence, one gram-ion of zinc is poorer in energy, by this amount, than one gram-atom of neutral metallic zinc.

If, following Ostwald, each positive electric charge on a kation is represented by a dot ($\cdot$), and each negative charge on an anion by a dash ($'$), then ionisation processes may be expressed in thermo-chemical equations as follows :—

$$Cu = Cu^{\cdot\cdot} - 17{,}500 \text{ cals.}$$
$$Zn = Zn^{\cdot\cdot} + 32{,}600 \quad ,,$$
$$Cl = Cl' + 40{,}100 \quad ,,$$

CHAP. V.] THE DISSOCIATION THEORY OF ARRHENIUS. 65

Various other heats of ionisation can be determined in the same way as j was found for zinc, from the heat of ionisation of copper and certain thermo-chemical data. These values are of especial interest. Some of those that have been determined by Ostwald are given here; they are all referred to the quantities of ions which are endowed with a single valency :—

K = + 61,000 cals.	Pb = − 500 cals
Al = + 39,200 ,,	H = − 800 ,,
Zn = + 16,300 ,,	Cu = − 8800 ,,
Fe = + 10,000 ,,	Hg = − 20,500 ,,
(ferro-ion)	Ag = − 26,200 ,,
Sn = + 1000 ,,	
Cl = + 40,100 ,,	

Although these numbers cannot be accepted as quite accurate, because of the difficulties in determining the values of w and E [in the equation $j = w - E$; see p. 64], nevertheless they show that the process of ionising an atom is accompanied sometimes by a gain, and sometimes by a loss, of energy; and that those elements which show greater chemical activity are poorer in energy in the ionic state, whereas energy has to be added from without to the less chemically active elements in order to ionise their atoms. Conversely, the neutralisation of the ions of the former elements can be accomplished only by the expenditure of much energy, whereas the ions of the latter elements readily separate from their solutions. [All the metals which are easily ionised show positive heats of ionisation; the others show negative heats of ionisation.]*

Especial stress should be laid on the fact that the process of ionisation does not necessarily involve a consumption of

---

* I have intercalated this sentence from Ostwald's paper referred to on p. 63.—[TR.]

5

energy; all that can be said in general terms is that *the energy inherent in the atoms undergoes a transformation during the process of ionisation*. While a portion is changed into electrical energy, another portion may pass out of the system; or energy may be taken up from without, whether in the form of heat, or in the form of light energy, both of which forms are changed into electrical energy during processes of ionisation. The total quantity of energy in an element when it is ionised is greater or less than before ionisation in accordance with the general chemical character of the element.

It was not at first easy for the new theory of electrolytic dissociation to hold its ground, and although the number of its adherents has increased very rapidly, thanks to the assiduous labours of its champions, there are still both physicists and chemists who cannot see their way to accept the existence of free portions of molecules carrying electric charges.

Those physicists, for the most part, regard the way of working of the electrolysing current from the point of view of the older theory of Grotthuss, which dates from the year 1805. This theory represented the work of the current as consisting in the ordering of the molecules in rows, and the separation of the ions, at the electrodes, from this association of molecules; and it was supposed that this explained the use of the electrical energy and its transformation into chemical energy. As early as 1857 Clausius (*Mechanische Behandlung der Elektricität* [1879], Part VI.) urged against this view the objection that, if the hypothesis were correct, a solution of an electrolyte could not act as a conductor until the current-energy (volt × ampère) had become sufficient to effect the decomposition of the molecules; and that from this moment, which would be marked by the sudden deviation of a galvanometer included in the circuit, very many molecules must be decomposed all at once. As a matter of fact, a current of the minimum number of ampères is able to accomplish electrolysis, provided the electrodes consist of the same metal as the kation of the electrolyte, and provided the tension that prevails at the electrodes is just sufficient to overcome the electromotive force acting in the opposite direction, which force is generally small, and is dependent both on the material of the electrodes and on the character of the ions, which give up their charges more or less readily. When these conditions are fulfilled, the needle of the galvanometer begins to move. The deviation increases quite gradually as the electromotive force of the current increases, and as, in consequence, the quantity of current passing through the solution increases, the quantities of the ions that are separated also become greater in accordance with the law of Faraday. Experience shows that conductors of the second class obey Ohm's law absolutely; but this could not be the case if the hypothesis of Grotthuss were accepted [without any modifications or additions].

If the energy of the current were really employed in splitting the molecules of the electrolyte, then those electrolytes whose ions are held together by weak affinities, in the chemical sense of that term, are just the electrolytes which ought to exhibit large conductivities. Experience goes against this conclusion also; for a solution of mercuric chloride conducts much worse (because of the smaller dissociation) than a solution of potassium chloride, and when potassium-silver cyanide is electrolysed, the potassium, which must be more firmly held than the silver, goes to the kathode, while the silver travels with the cyanogen to the anode. The dissociation theory is free from those objections which apply to the older theory; it takes the facts of electrolysis into account in a complete way, as has appeared from the examination of the theory made in preceding paragraphs.

Those chemists who deny the existence of free ions, because the chemical behaviours of these ions are said to be different from the behaviour of the neutral atoms, are to be reminded that *the quantities of energy associated with the ions are different from the quantities that are associated with the free elements, and that the ions must, therefore, differ qualitatively from the free elements.*

That the K· in a solution of potassium chloride does not react with the water, that hydrogen is not given off from this solution, and that the Cl' is odourless, depend either on the chemical energies of the ions being different in degree from the chemical energies of the free elements, or on the stoppage, by the electrical charges on the atoms, of the exhibition of the ordinary chemical effects of the energies of these atoms. Zinc dissolves in hydrochloric acid; but when it is charged negatively it does not dissolve. It is customary to explain the different degrees of readiness to enter into reaction that are shown by allotropic forms of such elements as phosphorus, oxygen, and carbon, by the supposition, which is well grounded, of each allotrope having a different energy-content from that of the others.*

Far from contradicting the facts of chemistry, the dissociation theory is in a position to render intelligible very many chemical processes that have remained obscure hitherto. Why the metals easily liberate hydrogen from the mineral acids while they are indifferent towards hydrocarbons; why the hydroxyl groups of the caustic alkalis readily split off

---

* Ordinary phosphorus (31 grams) = red phosphorus + 28,246 gram-calories.
  Ozone (48 grams) = ordinary oxygen + 36,200 ,, ,,
  Amorphous carbon (12 grams) = diamond + 3720 ,, ,,
  ,, ,, ,, = graphite + 3400 ,, ,,

when the alkalis react with salts of heavy metals, whereas the same groups are not removed from alcohols by similar reactions—for instance, from glycerin, which does not produce a precipitate in copper sulphate solution,—these reactions cannot be understood if it is assumed that the parts of the molecules of the acids and bases mentioned are more firmly held together than the parts of the molecules of the organic compounds. The great readiness to enter into reactions of those inorganic compounds which are electrolytes, the rapidity wherewith they accomplish their interactions compared with the slowness of reactivity shown by the non-electrolytes, to wit the carbon compounds, these facts were made intelligible for the first time by the dissociation theory (*see* Ostwald, *Zeit. für physikal. Chemie*, 2, p. 270 [1888]). The substances which are the most chemically active in solution are just those substances whose molecules are dissociated ; and this is a statement of which not analytical chemistry only, but synthetical chemistry also, especially the chemistry of carbon compounds, makes the most far-reaching applications. Moreover, the final results towards which the chemical processes that occur with the aid of electrolytes tend are at once made manifest by the dissociation theory. For the minute particles exchange ion for ion, and this the more quickly the greater is the mobility of the ions, and the more completely the dissociation has advanced. The dissociation-coefficient of Ostwald, a [see p. 59, where it is proposed to call a the *ionisation-coefficient*], gains a wider meaning as it becomes also the coefficient of activity in chemical reactions, and it will perhaps point the way by which measurements of chemical affinity may be made.

If the saponification of ethereal salts proceeds at the same rate whatever base or acid be used, provided only that the degrees of dissociation of the acids or bases employed are the same, this supposes the presence of free OH', or free $H·$, and therefore a dissociation of the molecules of the base, or of the acid. This process of saponification is carried on at the same rate by solutions of all bases and acids of equal molecular concentration, provided that $a = 1$, just as all very much diluted solutions of acids, containing the same number of gram-molecules, invert cane-sugar at the same rate.

It is evident that the reagents which serve to detect an element when that element is in the ionic condition, will no longer be serviceable when the element in question has combined with others to form a compound ion. For instance, the chlorine in the ClO_3' ion of potassium chlorate cannot be detected by a solution of a silver salt. Further, the facts that the atom of iron in the potassium salts of ferro- and ferricyanic acid is not precipitated by ammonium sulphide, and that copper is not precipitated from solutions of its salts by caustic soda when tartaric acid is present, find a reasonable explanation in the statement that the conditions of normal reactions are not fulfilled in these cases, because the metals iron and copper do not form independent ions, but are constituents of complex ions, in the compounds mentioned.

In connection with these reactions a sharper limitation may be arrived at of the conception of *a double salt* and *a salt of a complex acid*. When a solution of a double salt is electrolysed, both metals separate as kations; whereas the metal of a complex anion travels with the rest of that anion to the anode.

Attention should still be directed to the facts, that both the thermal neutrality that is observed when solutions of salts which do not react to produce precipitates are mixed (for instance, KCl and $NaNO_3$, or $AgNO_3$ and $CuSO_4$), and also the phenomenon of the equality of the heats of neutralisation [of dilute solutions of strong acids, or strong bases], are explained by the ionisation theory. The quantity of heat set free by the neutralisation of a [strong] acid in solution by a [strong] base also in solution to form a salt which remains in solution depends exclusively [according to the theory] on the combination of the H ions of the acid with the OH ions of the base, and it must, therefore, be independent both of the anion of the acid and of the metal of the base ; moreover, it must always have the same value,

$$H\cdot + OH' = H_2O + 13,700 \text{ cals. at } 20° \text{ [for a monobasic acid]},$$

unless other changes of energy occur in consequence of incomplete dissociation (see Arrhenius, *Zeit. für physik. Chemie*, **4**, p. 96 [1889]). [Compare Ostwald's article, "Electrical Methods," in Watts' *Dictionary of Chemistry*, new edition, vol. iv., pp. 189, 190.]

Ostwald (*Zeit. für physik. Chemie*, **2**, p. 271, and **3**, p. 120 [1888-89]) has described experiments intended to furnish direct evidence of the existence of free ions ; but only one of the experiments appears to me to be capable of being readily performed, by the arrangement which is described in the following paragraph.

A glass tube, 40 centims. long and 1 centim. diameter (RR, fig. 19), is bent at right angles near the ends, and two pieces of tubing, each about the size of a test-tube, are fused on to it, as shown in the figure. One of these tubes is tightly closed by a good cork, through which passes a rod of very pure amalgamated zinc, a. A piece of platinum, k, is fused through the walls of the other tube near the bend; the apparatus is filled with diluted sulphuric acid, and a cork is then inserted into this tube, which cork carries a narrow tube, bent twice at right angles, and containing coloured water, M. The narrow tube serves as a manometer. If the rod of zinc is now connected with the positive pole, and the platinum wire with the negative pole, of a battery of five accumulators, hydrogen instantly appears on the platinum wire, and the pressure of this hydrogen causes the liquid to rise in the manometer.

Now, if it was necessary for the current first of all to decompose the molecules of sulphuric acid, then the two atoms of

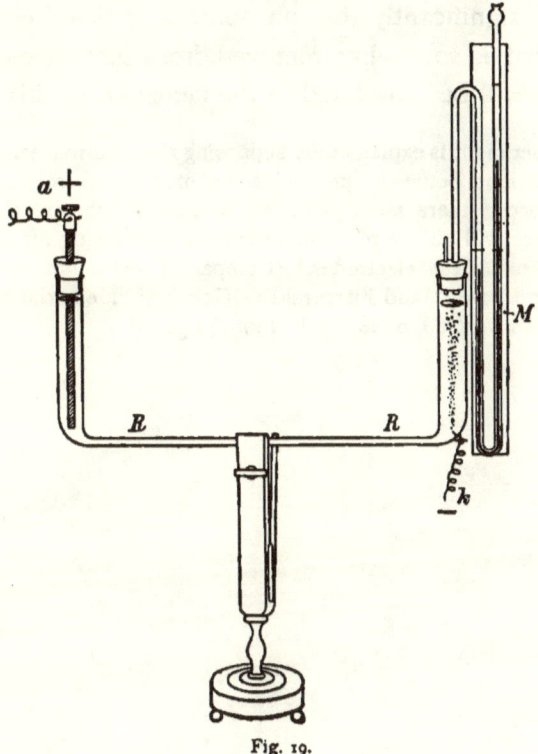

Fig. 19.

hydrogen from which the zinc had withdrawn the SO_4 radicle must have travelled to the platinum through the tube, which is 40 centims. long. Other experiments have shown that this passage of the hydrogen would occupy several hours. But as hydrogen appeared on the platinum at the moment of closing the circuit, free hydrogen ions must have been already present in the neighbourhood of the platinum, and these must have escaped as gas after their electrical charges had been neutralised.*

The dissociation theory seems, then, not only to be vindicated against all objections, but it is demanded for the explanation of diverse processes. The justness of the theory is exhibited yet more significantly by phenomena which belong to a domain that is somewhat removed from that of electrolysis, and which will be considered in the next part of this book.

* This experiment is explained by supposing the occurrence of a frequent exchange of ions between the molecules of the electrolyte. If such exchanges occur, there must probably be periods whereat the slightest directive force will suffice to precipitate the ions, that are for the moment nearly free, on to the electrodes. (Compare Lodge's *Modern Views of Electricity*, pp. 72-87 ; and Fitzgerald's "Helmholtz Memorial Lecture" in *Chem. Soc. Journal*, 69, p. 885 [July 1896].)—[TR.]

PART II.

THE THEORY OF SOLUTIONS OF VAN'T HOFF.

A THEORY of solutions was promulgated by van't Hoff simultaneously with the announcement of Arrhenius' theory of electrolytic dissociation. This theory has achieved extraordinary success in the short time during which it has been known, that is, since the second half of the last decade. The theory has made it possible to bind together, and to establish on a theoretical foundation, a number of phenomena which could not before be brought into line with one another. It has also been of great service in the practical work of chemistry, inasmuch as it has called into existence extremely valuable methods for determining molecular weights.

The points that will be especially attended to in this part are, the connections between van't Hoff's theory and the subject of electrolytic dissociation—for that subject has been greatly aided by the theory—and the elucidation of the conception of osmotic pressure. As the theory of van't Hoff is directly connected with Nernst's theory of the origin of the current, which will be considered in the last part of this book, it is necessary to go somewhat fully into the work of van't Hoff, although it may seem at first sight as if we were dealing with matters that have nothing to do with electrochemistry.

CHAPTER I.

OSMOTIC PRESSURE.

WHEN a body does not react chemically on a liquid wherein it is soluble, we are accustomed to regard the process of dissolution as purely physical and the solution as a molecular mixture. The changes of volume-energy and thermal energy which accompany the process of dissolution are positive or negative according to the state of aggregation of the substance to be dissolved, and these changes generally proceed in the same direction when a concentrated solution is diluted with the solvent. But when the dilution has reached a definite limit, these changes of energy can no longer be observed: the volume of the dissolved substance is then small compared with that of the solvent. The following considerations apply in the main to solutions whose concentrations approach that limit; although more concentrated solutions serve better for certain demonstrations, because they produce more marked effects and produce them more rapidly.

If a more dilute solution of a coloured salt—for instance, an aqueous solution of copper sulphate or potassium bichromate—is poured carefully, with the aid of a cork disc on the end of a glass rod, over a more concentrated solution of the same salt, it is observed that the concentration of the former solution gradually increases, while that of the latter decreases. The molecules of the dissolved substance diffuse from places of greater, to places of less, concentration, until the liquid is homogeneous throughout. The force which produces the diffusion corresponds with the gaseous pressure which compels a volume of gas to occupy a larger space when the chance is given to it to do so. For just as the gaseous pressure drives the gas-molecules to the new limits, so are the molecules of the dissolved body urged to distribute themselves equally in the solvent which has been added.

The following experiment serves to demonstrate the actual existence of such a pressure during the process of diffusion. A cylinder of about 100 c.c. capacity is filled to the brim with a concentrated syrupy solution of sugar, and the cylinder is then covered, air-tight, with a piece of animal membrane. When the cylinder is then sunk upright in a vessel full of water, the membrane begins to swell outwards in a cup-like form, and after some hours it reaches a height of a couple of centimetres. This phenomenon evidently depends on the striving of the sugar molecules to pass into the water outside the cylinder; but as they cannot do this because of the membrane, they stretch the membrane, and water flows into the space that has been formed in the cylindrical cell by this stretching. That the tension of the membrane is very considerable is shown by removing the cylinder from the water and puncturing the membrane with a fine needle; a stream of liquid is forced upwards through the small hole to a height of about 10 centimetres.

The process observed in this experiment is called *osmose* (ὠσμός = impulsion), and the force wherewith the molecules of the dissolved substance press on the membrane is called *osmotic pressure*. This pressure is found to be the greater the more concentrated is the solution.

In order to discover the more exact relations between osmotic pressure and concentration it is necessary to make use of membranes which are completely *semipermeable*, that is, which can be passed through by the molecules of the solvent, but not by those of the dissolved, body. This condition is not completely fulfilled by an animal membrane; for the presence of sugar can be detected in the water wherein the cylinder stood for some hours in the experiment that has just been described.* Perfectly semipermeable membranes are known, but they are few in number.

If the epidermal cells from the underside of the mid-rib of a leaf of *Tradescantia discolor* are floated in a 10 per cent. solution of nitre, the protoplasmic contents of the cells may be seen, under the microscope, to become detached from the cell-walls, and to contract, while the salt solution occupies the space between the protoplasm and the walls of the cells. H. de Vries, who studied osmose in plant-cells in 1884, called this phenomenon *plasmolysis*. The delicate pellicle which surrounds the protoplasmic contents of the cells acts as a semipermeable membrane in this experiment. This membrane allows the passage of water from the contents of the cell when the concentration of the salt solution is a very little greater than that of the cell-sap. When de Vries had ascertained the concentrations of those aqueous solutions of different substances which just brought about the plasmolytic condition in plant-cells, he found that these solutions were *equimolecular*, that is to say, he found that they contained quantities of the dissolved substances proportional to the molecular weights of those substances. *Equimolecular solutions, then, exhibit equal osmotic pressures; they are isotonic* (ἰσότονος = equally strained). *Hence the magnitude of the*

---

* The sugar is detected by boiling a small quantity of the water wherein the cylinder stood with a trace of dilute sulphuric acid, and then adding an excess of a hot *Fehling's solution* (10 grams copper tartrate + 500 grams water + 400 grams pure caustic soda), when a red precipitate of cuprous oxide is produced.

osmotic pressure is conditioned only by the number of the molecules in solution.

Valuable although this result is, yet direct measurements of the osmotic pressure of a solution of determinate concentration cannot be made by the plasmolytic method. To effect such measurements an apparatus is required the essential part of which is an artificially prepared semipermeable membrane. The choice of a medium for making such membranes is extremely limited, as very few suitable substances have been discovered, and the membranes made from these substances have possessed the desired property only in respect to the solutions of a small number of substances. But notwithstanding these difficulties, the few investigations of osmotic phenomena that have been conducted with artificial membranes have led to important results, and have incited further speculations.

The earliest experiments were made by Traube (*Archiv für Anat. und Physiol.*, 1867, p. 87). One shall be described here. A mixture was made of 5 c.c. of a 2·8 per cent. solution of copper acetate and 0·5 c.c. of a 10 per cent. solution of barium chloride; a glass tube about 5 mm. diameter was partially filled, by suction, with this mixture, and the upper end of the tube was closed by a piece of caoutchouc tubing and a pinchcock. The tube was then immersed in a 2·4 per cent. solution of potassium ferrocyanide (equimolecular with the solution of copper acetate), until the levels of the liquids were equal. A gelatinous precipitate of copper ferrocyanide soon formed at the lower end of the tube, and covered the opening of the tube like a skin. As this precipitated membrane allows water, but hardly any barium chloride, to pass through it, water was pressed into, the glass tube, and the membrane was extended outwards, like a bubble, by the osmotic pressure exerted by the solution of barium chloride.

Traube supposed that the semipermeability of a membrane of copper ferrocyanide is due to the membrane acting like a sieve, the meshes of which allow the smaller molecules of water to pass, but stop the larger molecules of the dissolved substance. Tamann (*Zeit. für physikal. Chemie*, **10**, p. 255 [1892]) has found, however, that these membranes are quite impermeable by solutions of the chlorides and nitrates of calcium and magnesium, but only partially impermeable by the chloride and nitrate of barium; hence he thinks that Traube's hypothesis is wrong. According to Tamann, freshly precipitated copper ferrocyanide is a hydrated compound which is able to act as a solvent towards certain substances; and the molecules of all those substances which dissolve in this compound are capable of passing through the membrane, while those substances which are insoluble in the compound are held back.

In order to carry out osmotic investigations on a larger scale with the precipitated membrane that has been described, it was still necessary to give to the membrane the proper ability to withstand resistance. This was done by Pfeffer (*Osmotische Untersuchungen*, Leipzig, 1877), by forming the membrane in the walls of a porous clay pot. The preparation of a membrane that is sufficiently compact and cohesive is an operation attended with some difficulty. It succeeds better the smaller the porous pot. Pfeffer used pots 4·6 centims. high and 1·6 centims. diameter. He fastened a glass tube, which served as a stopper, in the neck of a pot wherein a membrane had been formed; and to the side-piece of this glass tube he affixed a mercury manometer with the free limb closed; he then completely filled the apparatus with the liquid to be examined, closed it (*see* figure 20), and immersed it in a large quantity of water. The mercury gradually rose in the manometer tube, but several weeks elapsed before the maximum point was reached. When the maximum had been attained, then the pressure of the air in the closed limb of the manometer was in equilibrium with the osmotic pressure; and in this way Pfeffer was able to measure the osmotic pressure in atmospheres, and so to state the force wherewith the molecules of the dissolved substance pressed on a portion of the surface of the wall of the pot equal to the cross-section of the manometer tube.

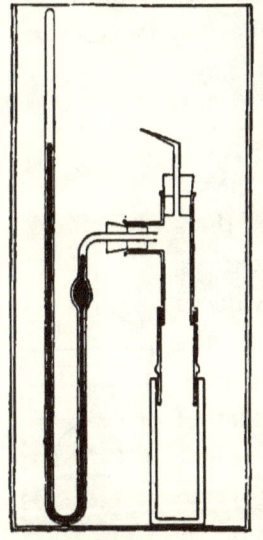

Fig. 20.

Because of their great importance, these experiments have been repeated from time to time with several changes in their arrangement. The most important change has consisted in making the precipitated membrane capable of resisting pressures of several atmospheres. There were reasons for thinking that more perfect semipermeability would thus be attained, and that it would be possible to extend the investigations to a greater number of substances. Reference should be made in this respect especially to the work of Tamann (*Zeit. für physikal. Chemie*, 9, p. 97 [1891]). But, after all, we have not as yet got much nearer the goal. Good results are to be expected by the use of Pringsheim's method of forming a very durable copper ferrocyanide membrane on a substratum of glycerin.

If it is desired to examine the phenomena of osmotic pressure merely qualitatively, a bell-shaped glass vessel of 225 c.c. capacity may be employed (*see* figure 21); the lower opening of this vessel, which is 7 centims. diameter, is closed by a porous plate formed by sawing off the bottom of a porous cell and filing it to the proper size; this porous plate must be fastened securely by means of sealing-wax. After the glass vessel has stood in boiling water for an hour (that it may be thoroughly saturated with water), it is filled with a 3 per cent. solution of potassium ferrocyanide, and sunk in a vessel which contains a 3 per cent. solution of copper sulphate until the levels of the liquids are equal. If the porous plate fits firmly, the brown precipitate of copper ferrocyanide will not be seen either in the glass vessel or in the liquid outside that vessel. The membrane that is produced acquires a sufficient thickness after three days.

The glass vessel is now prepared, once for all, for osmotic experiments; it is filled with a 50 per cent. solution of cane sugar, and closed with a cork through which passes a thermometer tube, r, of 1·3 mm. clear thickness, furnished with a scale. A little powdered alkali-blue is shaken into the bore of the thermometer tube before that tube is fixed in its place; this powder adheres to the sides of the tube, and makes the rise of the sugar solution very evident by the blue colour which it imparts to the liquid. If the apparatus is now fastened, by means of the cork k, in a vessel filled with water, C, the liquid in the thermometer tube rises at the average rate of 1 mm. per minute. After the experiment is finished, even if it proceeds for so long as five hours, the water in C shows merely a trace of sugar when tested with Fehling's solution.

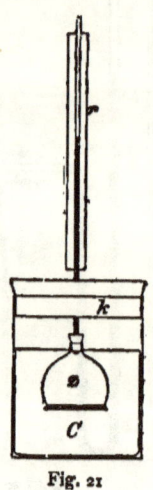

Fig. 21

Satisfactory results are obtained by the use of a cylindrical Pukall's filtering cell, the walls of which are made of sufficiently hard, porous porcelain.* A cell of this kind is got ready for

* These are to be obtained at the Royal Porcelain Manufactory, Berlin, for 0·75 mark apiece.

use in the following way. It is thoroughly soaked with water, by repeatedly evacuating under the air-pump [and then immersing in water]. It is then filled with a 3 per cent. solution of potassium ferrocyanide solution, closed with a cork which must not be set more than one centim. into the vessel, and which carries a piece of glass tubing open at both ends, and immersed in a 3 per cent. solution of copper sulphate. After seven days a sufficiently firm membrane of copper ferrocyanide has formed somewhere about the middle of the wall of the cell. For use in a demonstration, the cell is filled with a 50 per cent. solution of cane sugar, and it is then closed by a caoutchouc cork set as far as possible into the cell. The cork carries a ⊢-shaped glass tube, similar to that shown in the apparatus in figure 22; the side-piece of this tube, T, is connected, by means of a caoutchouc cork, with the manometer tube, M; this tube, which is furnished with a scale and contains indigo solution in both limbs, has a bulb blown on it, as shown at k, and the longer limb of it carries a little funnel, t. The tube S is now filled with the sugar solution, and closed firmly by the cork p. In order to remove any small air-bubbles that may be formed, a very thin glass tube, r, is pushed through p, and the capillary end of this tube is melted in the flame of a Bunsen burner. If the cell is now sunk over the rim in a cylinder filled with water, the indigo solution in the manometer rises at about the rate of 10 mm. per minute, if the bore of the tube is 0·79 mm. The osmotic taking up of water

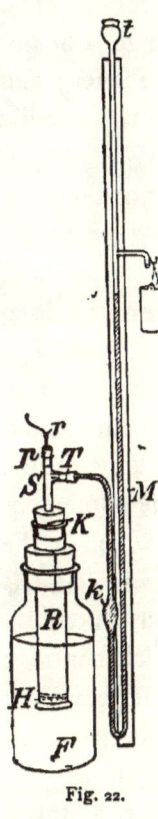

Fig. 22.

proceeds about four times as quickly as in the preceding experiment.

A Pukall's cell provided with a precipitated membrane seems to be suitable for quantitative, as well as for qualitative, osmotic experiments. When a 1 per cent. sugar solution was placed in the cell, and the manometer was filled with mercury, the mercury rose after some weeks to a height which agreed closely with the pressures found by Pfeffer; and not a trace of sugar could be detected in the water outside the cell.

The following experimental arrangement, recommended by Pfeffer (*loc. cit.*, p. 12), is not suited for measuring pressures, but it has the advantage that experiments can be carried out with it easily and quickly. A glass tube, 12 centims. long and 2·5 centims. wide (R, fig. 22), is used in place of a porous pot. The lower edge of this tube, having been flattened somewhat and ground, is brushed over with shellac, and a piece of parchment, H, is bound over it very firmly. The precipitated membrane is then formed in this parchment in the usual way. The rest of the arrangement, which is the same as that used in the experiment with a Pukall's cell, has been described above; the glass tube R is finally fixed, by means of a cork, in the neck of the bottle F, which contains water. It should be mentioned that 0·1 per cent. of potassium ferrocyanide is added to the sugar solution, and 0·09 per cent. (the equivalent quantity) of copper nitrate to the water, for the purpose of repairing damage done to the precipitated membrane during the experiment. The rate at which the liquid rises in the manometer varies to some extent with the thickness of the precipitated membrane. If the reaction of solutions of copper sulphate and potassium ferrocyanide is allowed to proceed for three days, a membrane will be produced of such a thickness that the liquid in the manometer will rise at the rate of from 1 to 2 mm. per minute. If a piece of parchment only is used, the liquid will rise about twice as rapidly at first; but sugar passes into the water—this does not occur until after about a couple of hours if the parchment is provided with a precipitated membrane—and, moreover, the maximum rise in the manometer is smaller.

The more accurate measurements of the osmotic pressures of solutions, by Pfeffer, led to the two following laws: *The osmotic pressure of a solution is proportional to the concentration, and also to the absolute temperature.*

Let P be the osmotic pressure in atmospheres, c the percentage contents of the solution, t the temperature (centigrade), T the absolute temperature, and a a constant dependent on the molecular weight of the substance in solution, and specifying the osmotic pressure at 0° and a concentration of 1 per cent.; then

$$P = a \cdot c\,(1 + 0\cdot00366\,t) = a \cdot c\frac{T}{273}.$$

According to Pfeffer, $a = 0\cdot649$ atmosphere for cane sugar.

The first to point out the analogy between these laws and the gaseous laws of Boyle and Gay-Lussac was van't Hoff (*Zeit. für physikal. Chemie*, 1, p. 481 [1887]).

The two gaseous laws are generally expressed together by the equation

$$pv = p_0 v_0\,(1 + 0\cdot00366\,t) = p_0 v_0 \frac{T}{273};$$

where p_0 is the gaseous pressure equal to 760 mm. of mercury, v_0 is the volume at 0°, and p and v are the observed pressure and the corresponding volume at $t°$. It is customary to employ a more profitable form of the equation in general chemistry. As both laws hold for all ideal gases, independently of their chemical compositions, Avogadro, as is well known, made the following statement: *Equal volumes of all gases, measured under the same conditions of temperature and pressure, contain equal numbers of molecules.* Not only has this statement been confirmed by numberless determinations of molecular weights, but it has also been deduced directly from thermodynamical principles. One litre of hydrogen measured at 0° and 76 centims. mercury-pressure weighs 0·08956 gram; hence one gram-molecule, that is two grams, of hydrogen occupies 22·38 litres at normal temperature and pressure. In accordance with Avogadro's law, the normal volume, that is to say, the volume occupied by a molecular weight in grams, of any gas must then be 22·38 litres. Now if, following Horstmann, the volume v_0 in the gaseous equation $pv = p_0 v_0 T / 273$ is put as the gram-molecular volume = 22·38 litres = 22,380 c.c., and if it is borne in mind that 76 c.c. of mercury weigh 1033·3 grams, and, moreover, if p is measured in grams and v in c.c., the equation takes the following simpler form :—

$$pv = \frac{1033\cdot3 \times 22{,}380 \times T}{273} = 84{,}700\, T \text{ gram-centims.}$$

Or, if p is measured in atmospheres and v in litres, the equation takes this form :—

$$pv = \frac{22\cdot38 \times T}{273} = 0\cdot0819\, T \text{ litre-atmospheres.}$$

(One litre-atmosphere = 1033·3 × 100 × 10 gram-centims.
   „   „   „   = 103·33 kilo.-decims.)

Or, finally, as 42,750 gram-centims. = 1 gram-calorie, the equation assumes this form :—

$$pv = \frac{84{,}700 \times T}{42{,}750} = 2\, T \text{ gram-cals.}$$

All these forms of the equation for pv are included in the general statement

$$pv = RT.$$

86 THE THEORY OF SOLUTIONS OF VAN'T HOFF. [PART II.

When the molecular volumes of gases, measured in the same units, are multiplied by the respective pressures, and the products are divided by the absolute temperature, a constant number is always obtained. The equation includes not only both gaseous laws, but Avogadro's law also. Very great use is made of it in stoichiometrical and thermodynamical calculations.

The following examples will serve to demonstrate the applicability of this gaseous equation.

I. To calculate the volume in litres occupied by 5 grams of hydrogen, the temperature being 27° and the pressure 72 centims. of mercury; the volume of 2 grams is first found,

$$v = \frac{0\cdot 0819 \cdot (273 + 27)}{72/76};$$

and from this it follows that the volume occupied by 5 grams is

$$= \frac{0\cdot 0819 \cdot (273 + 27)}{72/76} \times \frac{5}{2} = 64\cdot 837 \text{ litres.}$$

II. If the weight of 80 litres of carbon dioxide at 78 centims. pressure and 30° is to be found, the molecular volume (44 grams CO_2) at 78 centims. and 30° is determined,

$$v = \frac{0\cdot 0819 \cdot (273 + 30)}{78/76};$$

and then the weight that is sought, x, is arrived at by the proportion $v : 44 = 80 : x$; hence

$$x = \frac{78/76}{0\cdot 0819 \cdot (273 + 30)} \times 80 \times 44 = 145\cdot 5 \text{ grams.}$$

III. What amount of work is done by 1 kilo. of oxygen when it is heated to 100° at the atmospheric pressure?

One kilo. of oxygen contains $1000 / 32 = 31\cdot 25$ gram-molecules. Hence the work amounts to

$$84{,}700 \times 31\cdot25 \times 100 = 264{,}687{,}500 \text{ gram-centims.}$$
$$= 2646 \text{ kilogram-metres};$$
or $\quad 0\cdot0819 \times 31\cdot25 \times 100 = 256$ litre-atmospheres;
or $\quad\quad\quad 2 \times 31\cdot25 \times 100 = 6187$ gram-calories.

The gaseous equation proved of very signal service to van't Hoff in his further speculations concerning osmotic pressure. An extraordinarily close relation was found between osmotic pressure, P, and the gaseous pressure, p, when the equation $pv = 0\cdot0819\ T$ litre-atmospheres was applied to Pfeffer's results. The osmotic pressure $P = 0\cdot649$ atmos. which a 1 per cent. solution of cane sugar exerts at 0° was put by van't Hoff in place of p in the above equation. Now as 100 grams of water containing 1 gram of sugar in solution occupy 100·6 c.c., it follows that 1 gram-molecule (= 342 grams) of sugar is contained in 100·6 × 342 c.c. = 34·4 litres of a 1 per cent. solution of sugar. This volume was substituted for v in the foregoing equation; and van't Hoff then found R = 0·649 × 34·4 / 273 = 0·0818 litre-atmospheres, which is exactly the same value as that of the gaseous equation.

Then van't Hoff calculated the pressure of a gas whose volume is equimolecular with that of a 1 per cent. solution of cane sugar—that is, a gas containing 1 gram-molecule in 34·4 litres—at the temperatures whereat Pfeffer had determined the osmotic pressures of the sugar solution. The values of the two pressures are given in *Table V.*

TABLE V.

Temperature T.	Gaseous pressure, in atmos. calculated by formula $p = \frac{\cdot 0819}{34\cdot 4} T$	Osmotic pressure, P, in atmos. found by Pfeffer.
273·0	0·650	0·649
279·8	0·667	0·664
286·8	0·683	0·686
288·5	0·687	0·691

TABLE VI.

Percentage of sugar in solution.	V, the number of litres which contain 1 gram-mol. of sugar.	Gaseous pressure, calculated by $p = \dfrac{\cdot 0819 \times 288}{V}$ atmos.	Osmotic pressure, P, in atmos. found by Pfeffer.
1	34·4	0·687	0·691
2	17·3	1·349	1·337
4	8·8	2·667	2·739
6	5·9	3·956	4·046

Table VI. contains the osmotic pressures, P, of sugar solutions of different concentrations determined by Pfeffer at 15° ($T = 288$), and beside these are placed the calculated pressures of equimolecular volumes of a gas, p. The agreement of the magnitudes p and P, in both tables, is evident; and a similar agreement is shown for other substances besides cane sugar, as far as measurements of the osmotic pressures of solutions of such substances have gone.

The gaseous equation, $pv = \cdot 0819\, T$ litre-atmospheres, is, then, immediately applicable to solutions, when the osmotic pressure, P, is put in place of the gaseous pressure, p (in atmospheres), and for the gas-volume, v (in litres), is substituted the volume of the solution, V, that is, the number of litres that contain 1 gram-molecule of the substance. On the ground of this agreement, van't Hoff came to the conclusion that the osmotic pressure is equal to the gaseous pressure which would be observed were the solvent removed and the dissolved substance caused to occupy the same space as the solution, at the same temperature as that of the solution. This conclusion may be stated in the following words: *The molecules of a dissolved substance exert the same pressure against a semipermeable membrane, during osmotic processes, as they would exert against the walls of an ordinary vessel were they in the gaseous state at the same temperature and the same concentration.*

While thus extending the law of Avogadro to solutions, on the basis of Pfeffer's researches, van't Hoff also confirmed the law of de Vries, according to which osmotic pressure is not conditioned by the quality of the molecules of the dissolved substance, but only by the number of these molecules; and with these results he associated the conclusion that, under normal conditions, a substance which is not an electrolyte is separated by the process of dissolution into single molecules.

The energetics of solutions must be analogous to those of gases. Hence the same work must be done by the change of the volume-energy of a solution, measured by $P.dV$, as in the case of an equimolecular volume of gas under corresponding conditions of pressure and temperature. The measurement of osmotic pressures by Pfeffer's apparatus depends, on the whole, on this; inasmuch as the air is compressed in the closed limb, and the mercury is raised in the open limb, of the manometer, until the opposite pressure thus produced is equal to the osmotic pressure. In the first case there is merely a shifting of the volume-energy; in the second case there is a transformation of volume-energy into energy of position. A 1 per cent. sugar solution would produce a rise of 6·7 m. in the same solution in the open limb of a manometer. If the cross-section of this limb were 1 sq. centim., then the osmotic work done would be equal to $670 \times 981 \times 670/2 = 219,000$ ergs; for 1 c.c. opposes a resistance to the elevation of the liquid equal to 981 dynes, and the mean distance is $670/2$ centims. If a lateral overflow-tube were attached to the open limb at a less height than that just mentioned, liquid would flow from this tube until the solution in the cell had become so dilute, by taking up water, that equilibrium was attained between the osmotic pressure of this liquid and the hydrostatic pressure of the column of liquid in the manometer. The apparatus might thus be made to work as a machine for raising water (*see* fig. 22, p. 82, where a small pail is shown on a side-tube attached to the manometer.)

The effect of osmotic pressure, and, consequently, the analogy between osmotic and gaseous pressure, will be rendered clearer by imagining the following arrangement proposed by van't Hoff. In a vertically placed glass tube, open above and closed below, is placed a partition, formed of a semipermeable membrane, which fits securely into the tube, while at the same time it is capable of free movement up and down in the tube. Water occupies the space above this partition, and a solution is placed in the space under it. Now, if the pressure of the water-column is greater than the osmotic pressure of the solution, the partition will sink, and the solution will become more concentrated. But if the pressure of the water-column is less than the osmotic pressure of the solution, the partition will rise, and the solution will take up water, and so become more dilute. In both cases the movement of the partition will continue until an osmotic pressure is established which is just compensated by the pressure of the column of water.

The readiness to do work of a solution, exactly like the same quality of a volume of a gas, is the greater the more the volume is decreased, that is, the more concentrated the solution is made. But this work which the osmotic pressure of a solution performs, if opportunity is afforded the solution to take up solvent through a semipermeable membrane, must be expended in order to restore the former degree of concentration. And this work which is expended in the concentration of a solution cannot be measured directly by osmotic experiments, because the arrangement of van't Hoff is impracticable in such cases. Measurements can, however, be made indirectly by the aid of the processes of boiling and freezing, which generally make it possible to obtain a more accurate estimation of the osmotic pressure (see Chapters III. and IV. of this part).

As regards the magnitude of osmotic pressure, the high values which this pressure attains even in dilute solutions are certainly surprising. In a 5 per cent. sugar solution at 0° the pressure amounts to $5 \times 0.649 = 3.2$ atmospheres; and if the osmotic pressure of a half-saturated solution of ammonia is calculated, it is found to be equal to 671 atmospheres at 0°; this number is obtained by the equation $0.033\ P = 0.0819 \times 273$ (·033 litre of the solution of ammonia contains one gram-molecule, 17 grams, of NH_3). That the vessels wherein such solutions are kept are not shattered is to be accounted for by the internal pressure, calculated in thousands of atmospheres, which holds together the separate particles of the solutions. The osmotic pressure could only gain the mastery, and eventually destroy the walls of the vessels, if these walls were semipermeable and the vessels were immersed in the solvent (see Ostwald, *Lehrbuch der allgemeinen Chemie*, vol. i., p. 673 [1891].

The question as to how osmotic pressure is brought about has been discussed but little. The establishment of a kinetic theory of solutions is a task that is reserved for the future.

CHAPTER II.

THE VAPOUR-PRESSURES OF SOLUTIONS.

It is easy to measure the maximum vapour-pressure of a liquid by Dalton's method. All that need be done is to fill a glass tube, about 80 centims. long and 1 centim. wide, with mercury, quite free from air, to reverse the tube in a bath of mercury, and to allow a little flask filled with the liquid to rise to the top of the mercury in the tube; the little flask should be of about 1 c.c. capacity, it must be quite filled with the liquid, and it should be closed by a loosely fitting glass stopper. When the flask comes into the barometric vacuum, the stopper will be forced out; a portion of the liquid will evaporate, and the pressure of the vapour will press down the mercury a definite number of millimetres, which number expresses the maximum vapour-pressure of the liquid at the temperature whereat the experiment is performed. The level of the mercury must not be read off until after about ten minutes, and it is necessary to let the walls of the tube get wetted sufficiently by repeatedly inclining the tube. The magnitude of the barometric vacuum is of no essential importance in the measurement of the vapour-pressure, as it is only the quantity of vapour that is influenced by this factor, and the position of the mercury is not affected by the quantity of vapour. The result is valueless, however, if the barometer tube is not perfectly free from air, or if the liquid under examination contains impurities.

CHAP. II.]   THE VAPOUR-PRESSURES OF SOLUTIONS.   93

The value of the maximum vapour-pressure of a liquid is a measure of the volatility of that liquid. The value for water at 16° is 13·5 mm., while for ether it is 374 mm. at the same temperature.

The depression of the mercury in the barometer tube is found to be smaller when the liquid contains some other substance dissolved in it. The laws which express the lowerings of the vapour-pressures of solutions were elucidated experimentally by Raoult (in Grenoble) in the later years of the last decade (*Zeit. für physikal. Chemie*, 2, p. 353 [1888]). The work of that investigator was more fruitful than that of his predecessors (Wüllner and Babo), because he began with the more volatile solvents, and he paid especial attention to the solutions of those indifferent substances which did not conduct the electric current, and which themselves exerted vapour-pressures that were extremely small. Working by the barometric method, Raoult arrived at the following statements :—

I. The relative depression of the vapour-pressure

$$\frac{p-p_1}{p}$$

is independent of the temperature, from 0° to 20°. In the formula given p is the vapour-pressure of the solvent alone, and p_1 is the vapour-pressure of the solution.

II. The relative depression of the vapour-pressure increases in proportion to the quantity of dissolved substance, provided the concentration is not made too large.

III. The molecular depression of the vapour-pressure is obtained by referring the relative depression to one gram-molecule of the substance dissolved in 100 grams of the solvent ; thus

$$\frac{p-p_1}{p} \cdot \frac{m}{l},$$

where l is the quantity of the dissolved substance, in grams, and m is the molecular weight of the substance.

The molecular depression of the vapour-pressure has a constant value for solutions of different substances in the same solvent; it is dependent, like the osmotic pressure, only on the number of molecules of the dissolved substance that is present in the solution.

IV. If the number of molecules of the substance and the number of molecules of the solvent are expressed by n and N respectively (calculated by dividing the quantities by weight of substance and solvent by the respective molecular weights), then the following equation holds good for all solvents:—

$$\frac{p-p_1}{p} = \frac{n}{N+n}.$$

That is, the relative depression of the vapour-pressure, for any solvent, is equal to the ratio of the number of molecules of the dissolved substance to the total number of molecules of substance and solvent present in the solution.

Passing over the more accurate determinations of vapour-pressures which were made by Raoult, with materials that were purified very carefully, and by the use of the cathetometer for reading the levels of the mercury, and with all the necessary corrections, we may illustrate the law of proportionality—the relative depression of the vapour-pressure of a solution is proportional to the quantity of dissolved substance—by the following experiment.

Four barometer tubes of equal length, R_1, R_2, R_3, and R_4, figure 23, provided with millimetre scales, are filled with pure mercury, and are inverted in the mercury bath W and secured at equal distances apart. If all the air-bubbles have been removed, the levels of the mercury must be the same in the four tubes. Three very small flasks of the shape shown at F (such flasks as are used in Hofmann's vapour-density method) are filled, one with pure ether, another with a solution of 12·2 grams benzoic acid in 100 grams ether, and the third with a solution of 24·4 grams of the same acid in 100 grams ether. The best way of filling the little flasks is to attach them to platinum wire, and to immerse them in the solutions contained in the flasks wherein these solutions have been prepared, and as soon as the little flasks are full, to close them at once with their stoppers. The little flasks are then thoroughly cleaned externally by ether, and they are allowed to rise, stoppered ends downwards, and in the order stated above, to the top of the mercury in the tubes R_2, R_3, and R_4; the fourth tube, R_1, serves to measure the pressure of the air. The levels of the mercury become constant after a time, and it is then seen that if the levels are supposed to be joined by a curve, the curve will be a straight line, as is demanded by the law of proportionality. Although the numbers which are read off on the millimetre scales are only approximately accurate, nevertheless they help towards an understanding of Raoult's law of proportionality. They may, therefore, find a place in *Table VII.* (pp. 96, 97) beside the data of other experiments and the calculations belonging thereto.

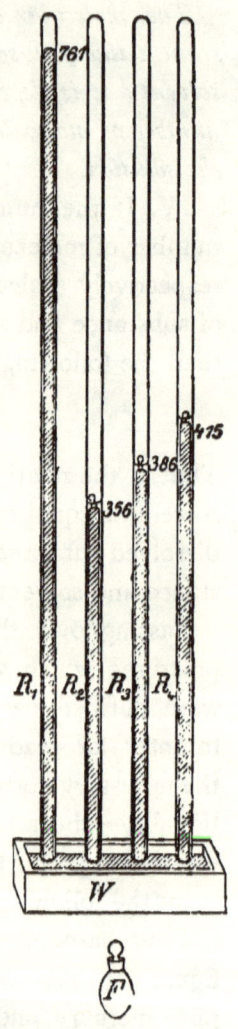

Fig. 23.

The numbers in columns 7 and 8 of *Table VII.* prove, approximately, the second and third laws of Raoult. The numbers in column 9 agree fairly with those in column 7, as

TABLE VII.

1	2	3	4	5	6	7
Solvent.	Dissolved substance.	Temperature.	Barometer.	p.	p_1.	$\frac{p-p_1}{p}$
	grams.	grams.		mm.	mm.	mm.	
I. Ether ; 100	Benzoic acid ; 12·2	17·75°	761	405	375	0·074
II. ,, ,,	,, ,, 24·4	17·75°	761	405	346	·1456
III. ,, ,,	Salicylic ,, 13·8	19·0°	755	420	388	·0762
IV. ,, ,,	,, ,, 27·6	19·0°	755	420	355	·1547
V. Benzene ,,	Naphthalene 12·8	21·0°	758	85·5	79	·0762

is required by the fourth law. The numbers in column 10 are obtained by dividing the values of the molecular depressions of the vapour-pressures by the molecular weights of the solvents (74 for ether, and 78 for benzene). These numbers give the relative depressions when 1 gram-molecule of substance is dissolved in 100 gram-molecules of solvent; they are nearly equal for both solvents to the value $1/(100+1) = 0·0099$.

The molecular weight, m, of the dissolved substance can be calculated by the help of Raoult's formula $(p-p_1)/p = n/N+n$; hence measurements of vapour-pressures serve to control the estimations of molecular weights. Let M be the molecular weight of the solvent, L the quantity of the solvent in grams, and l the quantity of the dissolved substance in grams, then

$$\frac{p-p_1}{p} = \frac{l/m}{L/M + l/m} = \frac{lM}{Lm + lM};$$

hence

$$m = M \cdot \frac{l}{L} \cdot \frac{p_1}{p - p_1}.$$

Column 11 of *Table VII.* contains the molecular weights of benzoic acid, salicylic acid, and naphthalene, calculated by this formula from the data given in the other columns.

TABLE VII.

8	9	10	11	12
			Molecular weight.	
			Calculated.	
$\frac{p-p_1}{p} \cdot \frac{m}{l}$	$\frac{n}{N+n}$	$\frac{1}{74\,(78)} \cdot \frac{p-p_1}{p} \cdot \frac{m}{l}$	$m = M \cdot \frac{l}{L} \cdot \frac{p_1}{p-p_1}$	Found by other methods.
0·740	0·0689	0·00999	113	122
·728	·1289	·00984	106	122
·762	·0689	·01030	124	138
·773	·1289	·01040	112	138
·762	·0723	·00977	121	128

Inasmuch as both osmotic pressure and relative depression of vapour-pressure are constant for equimolecular solutions in the same solvent, it may be assumed that these two magnitudes are connected causally. As a matter of fact, van't Hoff has deduced the one law from the other by a thermodynamical calculation.

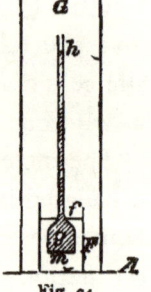

Fig. 24.

In a simpler way, Ostwald (*Lehrbuch der allgemeinen Chemie*, i., p. 728 [1891]), following Arrhenius, has arrived at Raoult's law from the law of osmotic pressure; and his deduction may be repeated here briefly. A small bell-shaped glass vessel, g (fig. 24), is closed beneath by a semi-permeable membrane, m, and is fitted with an upright tube 1 centim. wide; into the vessel is poured a solution of n gram-molecules of some substance dissolved in N gram-molecules of solvent, and the vessel is immersed, as far as f, in the pure solvent contained in the receptacle F. The apparatus is placed on a plate, A, and it is covered with a bell-jar, G, inside of which a vacuum is produced. The osmotic process causes the liquid to rise in the upright tube, say as far as h. The osmotic pressure is then given by the equation

$$P = \frac{nRT}{V}.$$

where $R = 84,700$, P being measured in grams and V in c.c. Expressing the molecular weight of the solvent by M, the weight of the solution is MN grams, as the weight of the dissolved substance may be neglected because of the small concentration of the solution; and, putting s as the specific gravity of the solution—and this must be very nearly equal to the specific gravity of the solvent—then V must be $= MN/s$ c.c. Consequently

$$P = \frac{nsRT}{MN},$$

Moreover, if the distance fh, measured in centims., is represented by H, then $P = Hs$; and hence

$$H = \frac{nRT}{MN}.$$

But if the apparatus is in equilibrium, the vapour-pressure of the solution p_1, at the point h, is equal to the vapour-pressure of the solvent p, decreased by the weight of a column of vapour of the length fh and 1 square centim. cross-section. Hence, if d represents the weight of 1 c.c. of vapour, $p_1 = p - Hd$, or $p - p_1 = Hd$. Now v c.c. of vapour weigh M grams, when v is the molecular volume of the solvent in the form of vapour. Hence 1 c.c. vapour weighs M/v grams; or, as the gaseous equation $pv = RT$ is applicable,

$$d = \frac{Mp}{RT}.$$

Finally, by inserting the values for H and d in the equation $p - p_1 = Hd$, we get

$$p - p_1 = \frac{nRT}{MN} \cdot \frac{Mp}{RT} = \frac{n}{N}p;$$

or

$$\frac{p - p_1}{p} = \frac{n}{N}.$$

CHAP. II.] THE VAPOUR-PRESSURES OF SOLUTIONS. 99

This is the equation found empirically by Raoult; provided that n is very small in reference to N, that is, provided the solution is very dilute.

For limited concentrations H has an extremely high value, and as d varies with H, the pressure π of the column of vapour of the height H cannot be put down, offhand, as $= Hd$. Rather is $\delta\pi = d \cdot \delta H$; or, as $d = M\pi/RT$, then

$$\frac{RT}{M} \cdot \frac{\delta\pi}{\pi} = \delta H.$$

Integrating this equation between 0 and H, we get

$$H = \frac{RT}{M} \log \frac{\pi_0}{\pi_H},$$

and in this $\pi_0 = p$ and $\pi_H = p_1$. But it has been found from the osmotic pressure that

$$H = \frac{nRT}{MN}.$$

Hence

$$\frac{RT}{M} \log \frac{p}{p_1} = \frac{nRT}{MN},$$

or

$$\log \frac{p}{p_1} = \frac{n}{N}.$$

But

$$\log \frac{p}{p_1} = \log \left(1 + \frac{p - p_1}{p_1}\right) = \frac{p - p_1}{p_1} - \frac{1}{2}\left(\frac{p - p_1}{p_1}\right)^2 + \ldots \frac{p - p_1}{p_1}.$$

Hence it follows that

$$\frac{p - p_1}{p_1} = \frac{n}{N},$$

or

$$\frac{p - p_1}{p_1} = \frac{n}{N + n}.$$

Thus the empirical formula of Raoult is placed on a theoretical foundation; and the formula $PV = RT$, on which that way of considering the matter rests, is again justified.

* Compare Poynting, *Phil. Mag.* [5], 42, p. 289 (October 1896), where the formula $\frac{p - p_1}{p_1} = \frac{n}{N}$ is arrived at on the hypothesis of aggregation of molecules of the solvent and the dissolved substance.—[TR.]

CHAPTER III.

BOILING POINTS AND FREEZING POINTS OF SOLUTIONS.

WHEN a solution boils, it is known that the solvent alone volatilises, provided the boiling point of the dissolved substance is about 130° higher than that of the solvent. When a solution is frozen, the solvent alone separates in the solid form if the solution is not too concentrated; this has been proved directly by Rüdorff with a solution of magnesium platinocyanide. Considering that the vapour-pressure of a solution is always less than the vapour-pressure of the solvent, it follows that a solution must boil at a higher temperature and freeze at a lower temperature than the pure solvent. For the vapour-pressure of the solution will not yet be able to overcome the pressure of the air at the boiling point of the solvent; to accomplish this, that is, to make it possible for it to boil, the solution must be raised to a higher temperature. Moreover, the solvent begins to freeze when the vapour-pressure of the solution is equal to that of the solid solvent; but this condition is not fulfilled until a temperature is attained which is lower than the freezing point of the pure solvent. Hence, as the vapour-pressure of a solution is most intimately connected with the increase of the boiling point and the lowering of the freezing point of the solution, it is to be expected that, in accordance with Raoult's law of vapour-pressure, these two quantities will increase proportionately to the concentration, and that they will have the same value for equimolecular solutions in the same solvent.

As a matter of fact, experiments have led to this conclusion; and it was Raoult who succeeded in solving this problem experimentally, especially by his investigation of organic compounds, and his statement of the results in molecular quantities of these compounds.

Inasmuch as the methods based on determinations of the boiling points and the freezing points of solutions give more accurate estimations of the molecular weights of the dissolved substances, and are more convenient to use, than methods dependent on measurements of vapour-pressures, apparatuses of great delicacy and accuracy have been constructed from time to time for the purpose of determining the boiling points of solutions. Those described by Beckmann are used most commonly (*Zeit. für physik.*

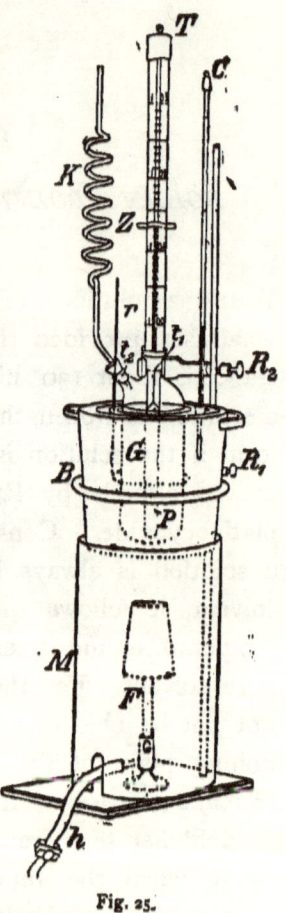

Fig. 25.

Chemie, **2**, p. 639; and **4**, p. 543). The apparatus represented in figure 25 is well adapted for determining the boiling points of aqueous solutions for demonstration-purposes. This apparatus is based on that of Beckmann; certain changes have been introduced, with the object of making the results visible from a distance. The vessel B is a glycerin bath of 1½ litres capacity; it is fixed on the ring of a stand, R_1, at a distance of 30 centims. above the flame of the burner F. A mantle of cardboard, M, protects the flame of the burner from draughts of air. The thermometer C is fixed in a hole in one of the rings of the bath B; this thermometer serves to regulate the temperature of the glycerin which is stirred from time to time by the stirrer r. The boiling vessel G is of cylindrical form; it has a capacity of 280 c.c., and it is furnished with a tubulus, t_1, of 2·5 centims. diameter, by which the vessel is supported in the clamp R_2, and also with another tubulus, t_2, of 1·5 centims. diameter. The thermometer T is fastened by a cork in the tubulus t_1. The temperature whereat the solution boils is read off on this thermometer, and is marked by the movable indicator Z. The bulb of the thermometer P, which is filled with mercury, is 10 centims. long, and has a diameter of 1·8 centims. The scale comprises only three degrees, from 100° to 103°; each degree covers a length of 7 centims., so that hundredths of a degree are indicated.* The condenser K is fixed in the tubulus t_2. This condenser is a tube with seven bends, each of which is 5 centims. diameter; above the lower end, which is cut off obliquely, is an opening whereby the vapour may enter the condenser without being hindered by the condensed water that flows back into the vessel G.

The following points are to be noted in using the apparatus. The glycerin must fill the vessel B nearly to the ring R_1; and the solution in G must stand at a level with this ring. The thermometer T is to be sunk so deeply that the level of the bulb coincides with that of the boiling liquid. It is

* These thermometers are to be obtained from Warmbrunn, Quilitz & Co., Berlin, at a price of 18 marks

better to make the solution and then to transfer it to the apparatus than to add the substance to be dissolved, through the tubulus t_2, to the solvent in the vessel; because only concentrated solutions show sufficiently large increases of boiling point; but the addition of a large quantity of substance to be dissolved considerably increases the volume of the liquid. In order to ensure regular boiling as far as possible, and also to prevent bumping caused by the alternation of overheating and sudden formation of vapour, it is advisable, either to throw little pieces of soapstone into the boiling-vessel, through t_2, just before the temperature of ebullition is reached, or to place a number of little capillary tubes, sealed at their upper ends, in the liquid before the heating is commenced. Moreover, to prevent undue delay in the process of boiling, the temperature of the glycerin should be at most 2° higher than the boiling point of the solution, which may be estimated beforehand within narrow limits. The temperature of the glycerin-bath can be kept constant by regulating the flame by means of the pinchcock h and moving the stirrer r frequently up and down; when the temperature of ebullition is nearly reached, the flame should be not more than from 3 to 4 centims. long. The liquid must be allowed to boil for about ten minutes before the temperature is read, in order that the large quantity of mercury in the thermometer may become heated equally; and the thermometer must be tapped repeatedly because of the inertia of the thread of mercury. Moreover, as the boiling point varies with variation of the barometer, and the thermometer changes somewhat in course of time, the boiling point of water must always be determined by a special experiment.

Table VIII. presents the results obtained with an aqueous solution of cane sugar; l shows the grams of sugar in 100 grams of water, b the height of the barometer in mm., t_0 the boiling point of water, t the boiling point of the solution, m the molecular weight of cane sugar (342), and S the molecular increase of boiling point referred to 100 grams of the solvent.

TABLE VIII.

1	2	3	4	5	6	7	8
l	b	t_0	t	$t - t_0$	$\frac{t - t_0}{l}$	$S = \frac{t - t_0}{l} \cdot m$	$m = S \frac{l}{t - t_0}$
grams.	mm.						
34·2	746	99·85°	100·35°	0·50°	0·01462	5·00°	342·0
51·3	746	99·85°	100·59°	0·74°	0·01442	4·93°	341·7
68·4	755	99·90°	100·95°	1·05°	0·01535	5·24°	341·3
85·5	755	99·90°	101·22°	1·32°	0·01544	5·28°	342·0

The numbers in column 6 show that the increase of the boiling point is constant for each 1 per cent. of sugar,—provided the solutions are not too concentrated; that is to say, the *boiling point increases in proportion to the concentration.* The values of S in column 7 indicate that the *boiling points of equimolecular solutions are increased to the same amount.* If the constant S has been found for a determinate solvent, by the use of a substance whose molecular weight is known, then the molecular weight m of another substance dissolved in the same solvent can be found by the equation $l : m = t - t_0 : S$. Column 8 contains the values of m, for sugar, calculated from the values found for l and $t - t_0$; the values of m agree very closely, and are nearly equal to the true value 342. If L grams of solvent are used, in place of 100 grams, then

$$m = 100\, S \frac{l}{L(t - t_0)}.$$

Troublesome though it is to determine the boiling points of aqueous solutions during a lecture, these boiling points have a very especial interest in connection with electrolysis (see Chapter V.). The experiments with cane sugar which have been described may be turned to further account, as it is possible to invert the cane sugar in accordance with the equation

$$C_{12}H_{22}O_{11} + H_2O = 2C_6H_{12}O_6$$

by adding a c.c. of hydrochloric acid containing potassium chloride to the boiling solution, and so to double the number of molecules in solution in less than a minute, and hence to bring about double the increase of the boiling point.

In making determinations of molecular weights by the boiling-point method, care must be taken to select solvents which are without chemical action on the dissolved substance.

The boiling points, and the values of S, of various solvents are as follows :—

	Water.	Alcohol.	Ether.	Acetic acid.	Ethyl acetate.	Chloroform.
$t_o =$	100°	78·3°	34·97°	118·1°	72·8°	61·2°
$S =$	5·2°	11·5°	21·1°	25·3°	26·1°	36·6°.

When one of these liquids other than water is used, thermometers with smaller bulbs may be employed, because of the higher values for S for these liquids. Experiments may therefore be conducted more rapidly with these liquids. The apparatus represented in figure 26, which is of more simple construction than that shown in figure 25, may be employed for the purposes of demonstration. The bulb of the boiling vessel G is of 200 c.c. capacity; its neck is 30 centims. long and 3 centims. wide. The vessel is placed in a basin made of asbestos lined with asbestos-wool, S, and it must be surrounded by a mantle, M, made of thick woollen cloth (as shown in the figure) to protect it from the effects of possible draughts of air. The cork k carries the thermometer T, and the two long serpentine condensers K_1 and K_2, between the bends of which wetted filter paper should be placed. The bulb of the thermometer (50 × 8 mm.) reaches so far into the boiling-vessel that the end of it touches the surface of the 100 grams of solvent placed in that vessel. The thread of mercury is 2 mm. broad; it may therefore be seen from a considerable distance. The scale begins at 30°; and from 30° to 110° it is 30 centims. in length, so that each degree is represented by nearly 4 mm., and tenths of a degree can be read off. Delay in boiling is avoided in the way already described (p. 103). An Argand-burner with a luminous flame, provided with a tall chimney and a regulating stopcock, is recommended for heating. The boiling point of the solvent is determined first of all, and the substance to be dissolved is then added through the neck of the boiling-vessel.

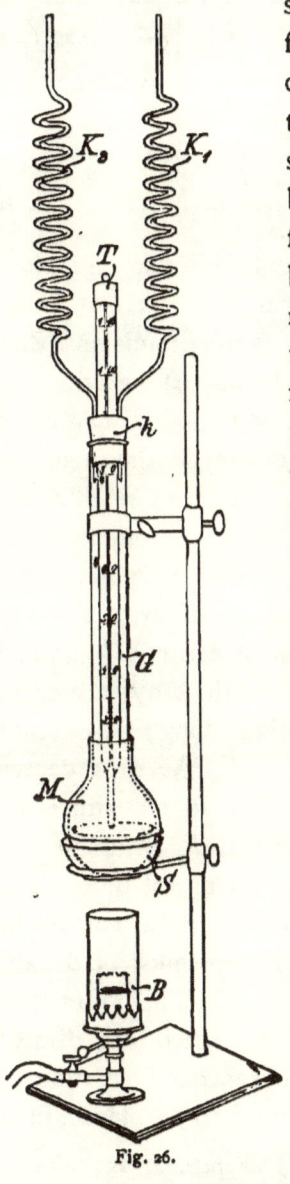

Fig. 26.

Table IX. contains the results of determinations of the boiling points of solutions of naphthalene (mol. wt. 128) in ethylacetate.

BOILING AND FREEZING POINTS OF SOLUTIONS.

TABLE IX.

l	t_0	t	$t-t_0$	$\dfrac{t-t_0}{t}$	$S = \dfrac{t-t_0}{t} \times m$	$m = 26 \cdot 1 \times \dfrac{l}{t-t_0}$
grams.						
3·2	72·8°	73·5°	0·7°	0·2188	28·00°	119·3
6·4	72·8°	74·1°	1·3°	0·2031	25·99°	128·5
9·6	72·8°	74·7°	1·9°	0·1979	25·33°	125·3

Decidedly fewer precautions are required in determinations of the freezing points of solutions than in determinations of their boiling points. Several years ago Ciamician (*Zeit. für physik. und chem. Unterricht*, **3**, p. 39) described a process by which lowerings of freezing points might be illustrated qualitatively by the use of an air-thermometer. But as the air-thermometer cannot be adjusted readily, the demonstration-thermometer T (fig. 27) is to be recommended.* This instrument is useful for many other purposes, such as determinations of the heats of solution of salts and the specific heats of metals. The bulb is 14 centims. long, and only 1·5 centim. wide; it is filled with amylic alcohol coloured blue. The scale is 60 centims. long; it extends over 75°, from 30° under to 45° above zero. As each degree is 8 mm. long, tenths of a degree can be read off with ease. This thermometer is placed in the vessel R (wherein the freezing is effected), which has the form of a test-tube. The level of the solution under examination must be coincident with that of the bulb of the thermometer, and must be distant 3 centims. from the rim of the vessel R. The tube R is fastened in the opening of the flask F by means of a disc of cork. The flask F is filled with a concentrated solution of calcium chloride in glycerin or alcohol, and is placed in a

* Made by glass-blower Stühl, Berlin N., Philippstrasse 22; price 12 marks.

cooling bath, K, of 5 to 6 litres capacity. To obtain approximately accurate results with this apparatus, the sensitiveness

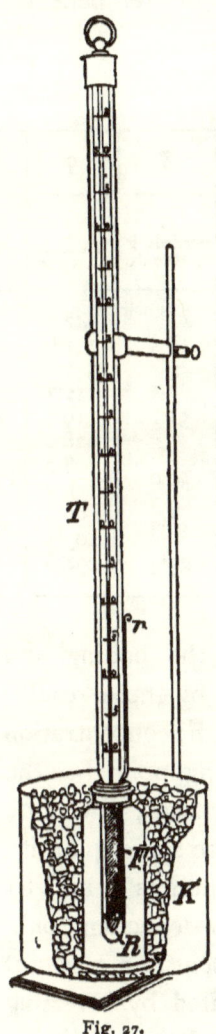

Fig. 27.

of which can only be somewhat small, it is necessary to work with solutions such that the freezing points to be observed are not very low, and the stirrer r (which is made of nickel) must be kept constantly in motion until the solvent separates in the solid form. The process of cooling must in no case be conducted very quickly; and it is therefore necessary that the temperature of the solution of calcium chloride should not be more than 2° or 3° lower than that of the freezing point of the solvent. This can be brought about easily, inasmuch as the proper temperature can be produced in the cooling bath K by using ice-water, or a mixture of ice with more or less common salt, according to the circumstances of the experiment.

The results of a few experiments made with solutions in water, and in benzene which froze at 5·5°, are given in *Table X*. In that table l is the quantity of substance dissolved in 100 grams of the solvent, θ is the freezing point of the solution, and θ_0 the freezing point of the solvent; G is the molecular depression of the freezing point, m (column 9) is the calculated molecular weight of the compound in solution, and m_0 is the actual molecular weight of that compound. The freezing points observed in these experiments are somewhat too

BOILING AND FREEZING POINTS OF SOLUTIONS.

low, because, independently of the defects of the method greater concentrations than are advisable were employed in order to obtain as large differences of temperature as possible.

TABLE X.

1	2	3	4	5	6	7	8	9	10
Solvent.	Dissolved compound.	l	θ	$\theta_0 - \theta$	$\dfrac{\theta_0-\theta}{l}$	$G = m \times \dfrac{\theta_0-\theta}{l}$	G according to Raoult.	$G \times \dfrac{l}{\theta_0-\theta}$	m_0
Water	Cane sugar	34·2	−1·8°	1·8°	0·053	18·13°		342·0	
,,	,, ,,	51·3	−2·8°	2·8°	0·054	18·45°	18·5	341·6	342·0
,,	,, ,,	68·4	−3·8°	3·8°	0·055	18·81°		342·0	
Benzene	Chloroform	11·9	+0·2°	5·3°	0·444	53·01°		112·7	
,,	,,	23·9	−4·5°	10·0°	0·419	50·01°	51·1	119·5	119·5
,,	,,	35·8	−8·5°	14·4°	0·391	46·72°		128·0	
,,	Naphthalene	12·8	+0·5°	5·0°	0·391	50·05°	50·0	128·0	128·0
,,	,,	25·6	−4·0°	9·5°	0·371	47·19°		134·7	
,,	Aniline	9·3	+0·8°	4·7°	0·505	46·96°	46·3	99·3	93·0
,,	,,	18·6	−3·5°	9·0°	0·484	45·01°		103·3	
,,	Benzoic acid	6·1	+4·2°	1·3°	0·213	25·98°	25·4	234·6	122·0

The analogy between the phenomena of the boiling and the freezing of solutions is shown very well by these results. The freezing point falls proportionally with concentration (column 6), and the depression of the freezing point has the mean value of 18·5° for water, and 49° for benzene, when one gram-molecule of substance is dissolved in the same quantity (100 grams) of either of these solvents (columns 7 and 8). That the depression of the freezing point is dependent only on the number of molecules of substance present in a constant quantity of the solvent may also be illustrated by inverting the 34·2 per cent. solution of cane sugar, and then bringing the solution again into the freezing apparatus, when θ will be found to be $= -3\cdot3°$. The numbers in columns 9 and 10 indicate the importance of determinations of the freezing points of solutions as means for finding molecular weights. The molecular weights are calculated by the equations:—

$$l : m = \theta_0 - \theta : G$$

or,

$$l : m = (\theta_0 - \theta) L/100 : G,$$

where L expresses the quantity of solvent used in the special experiment.

The results obtained by the boiling-point method and by the freezing-point method show differences from the numbers calculated from the theory; these differences cannot be referred altogether to the errors inherent in the methods. The numbers obtained by the use of Beckmann's apparatus often deviate very markedly from the theoretical values. The values of $(\theta_0 - \theta) / l$ decrease, as a rule, proportionally to concentrations, for solutions in benzene; this is explained by the assumption, which is well founded, that some of the molecules of the substance dissolved in benzene are associated into double molecules. Benzoic acid, especially, forms double molecules when dissolved in benzene. On the other hand the changes both of boiling and freezing points are greater in aqueous solutions of considerable concentration than the theory demands. In these cases the molecules of the dissolved substance must exert an attraction on the molecules of the solvent, which attraction hinders the volatilisation and freezing out of the latter molecules.

CHAPTER IV.

SUMMARY.

THE investigations made regarding dilute solutions have led, so far, to four laws of similar import. These laws are as follows: *Equimolecular solutions of any substances, prepared by using equal weights of the same solvent, exhibit equal osmotic pressures, equal relative depressions of vapour-pressure, equal raisings of boiling point, and equal lowerings of freezing point.* The experimental data allow the inversion of these laws. The magnitudes P, $(p - p_1)/p$, $t - t_0$, and $\theta_0 - \theta$ must, therefore, be proportional to one another for relatively equal concentrations, which are, however, generally small concentrations, and the factor of proportionality can be dependent only on the constants of the solvent. Van't Hoff (see Nernst's *Theoretical Chemistry* [English Ed.], pp. 126, 128) elucidated the relations between osmotic pressure and the three other magnitudes; and he established the following equations, in which M is the molecular weight, and s is the specific gravity, of the solvent, T_0 and T_0' are its absolute boiling point and freezing point respectively, w and w' are the latent heat of evaporation and the latent heat of fusion of the solvent per gram, and K, K', and K'' are collective constants depending on the solvents used.

I. $P = \dfrac{p - p_1}{p_1} \times \dfrac{819 s \cdot T}{M} = \dfrac{p - p_1}{p_1}$ TK atmos.

II. $P = (t - t_0) \times \dfrac{41\cdot37 \cdot s \cdot w}{T_0} = (t - t_0)$ K' atmos.

III. $P = (\theta_0 - \theta) \times \dfrac{41\cdot37 \cdot s \cdot w'}{T_0'} = (\theta_0 = \theta)$ K'' atmos.

Hence it must be possible to find one magnitude from the others. Values for P may be found by applying equations I. and II. to the experimental data which have been tabulated already. The results of doing this are to be seen in the two following tables:—

TABLE XI.

1	2	3	4	5	6	7	8	9	10	11
Solvent.	Dissolved substance.	l	V	$t-t_0$	s	w	T_0	K'	$P = \dfrac{}{(t-t_0)K'}$	$P = \dfrac{\cdot 0819\, T_0}{V}$
Water	Cane sugar	51·3	0·6667	0·74	0·959	536·4	373·0	57	42·18	45·82

TABLE XII.

1	2	3	4	5	6	7	8	9	10	11
Solvent.	Dissolved substance.	l	V	$\theta_0-\theta$	s	w'	T_0'	K''	$P = \dfrac{}{(\theta_0-\theta)K''}$	$P = \dfrac{\cdot 0819\, T_0'}{V}$
Water	Cane sugar	51·3	0·6667	2·8	0·9998	79·0	273·0	11·85	31·89	33·54
Benzene	Chloroform	11·9	1·1494	5·3	0·8700	29·5	278·5	3·76	19·93	19·84
,,	,,	23·9	0·5747	10·0	0·8700	29·1	278·5	3·76	37·61	33·68

In both tables P (in column 11) is calculated from the equation $PV = \cdot 0819\, T_0$, or $= \cdot 0819\, T_0'$, litre-atmospheres; where V is the number of litres which would contain one gram-molecule of the substance in solution. Considering that the solutions which were used in determining boiling points and freezing points were, relatively, very concentrated, it is seen that the values thus calculated for P agree very well with those in column 10 which are found by the formula of van't Hoff.

It is thus seen that the values found for P from determinations of boiling points and freezing points, which determinations can be accomplished much more accurately than measurements of osmotic pressure, satisfy the equation $PV = RT$, which corresponds with the gaseous equation; *hence the theory of van't Hoff is confirmed afresh by the methods of boiling and freezing points*. Moreover, testimony to the justness of this theory is borne by the fact that the same molecular weights are found for substances in solution from determinations of the boiling and freezing points of the solutions as those which are arrived at by measurements of vapour densities.

Another very convincing argument in favour of the theory of solutions has been given by van't Hoff, by his demonstration that the values arrived at for the molecular depressions of freezing points by purely thermodynamical considerations based on the gaseous equation are the same as those that are determined empirically. Let the following reversible process be imagined. A solution of n gram-molecules of a substance in a *very large number*, N, of gram-molecules of a solvent is at the temperature θ_0, and is cooled to θ. In the process of cooling so much heat is withdrawn that N/n gram-molecules of the solvent, that is, the quantity in which one gram-molecule of the substance is dissolved in the original solution, freeze out. Hence $N/n \cdot M\omega$ gram-calories will be set free, at θ, when ω represents the latent heat of fusion at θ. Let it be supposed that the quantity of the solvent which is frozen is removed from the solution, is heated to θ_0, and is melted at θ_0. In doing this $N/n \cdot M\omega'$ gram-calories will be taken up, when ω' is the latent heat of fusion of the solvent at θ_0. Now, as latent heats of fusion are smaller at lower temperatures, it is true that

$$\frac{N}{n} M\omega < \frac{N}{n} M\omega'.$$

If the remaining solution is brought, meanwhile, to θ_0, the positive quantity of heat which is added during all these processes is

$$\frac{N}{n} M (\omega' - \omega) \text{ gram-calories.}$$

The solution can perform work at the cost of this heat, if opportunity is given, by again taking up, osmotically, through a semipermeable membrane, N/n gram-molecules of the solvent. The initial condition of the cyclical process is thus attained.

This cyclical process can also be performed in the reversed direction. For one can think of N/n gram-molecules of the solvent being pressed out of the solution through a semipermeable membrane, at θ_0, by a piston, being then allowed to freeze at θ_0, then cooled along with the solution to θ, and being then brought back into the solution wherein it melts. If the whole system were now warmed to θ_0, the cyclical process would be completed, and the quantity of heat

$$\frac{N}{n} M (\omega' - \omega) \text{ gram-calories}$$

would become free in place of the osmotic work that was done on the solution. Moreover, as the formula

$$PV = 2T_0' \text{ gram-calories}$$

—where P is the difference between the osmotic pressures of the more concentrated and the more dilute solutions, and V is that volume of solvent which is osmotically taken up by, or is pressed out of, the solution—is applicable to the osmotic work, which in one case is done on the solution, and in the other case is performed by the solution, it follows that the equation

$$2T_0' : N/n \cdot M\omega' = \theta_0 - \theta : T_0'$$

holds good, when attention is paid only to the first cyclical process; for, in accordance with the second law of the dynamical theory of heat, when heat does work in a reversible cyclical process, the portion of this heat which is changed into work is related to the total quantity of heat added to the system as the fall of temperature is related to the absolute temperature whereat the heat is taken up by the system.

It follows from the foregoing equation that

$$\theta_0 - \theta = \frac{2n T_0'^2}{M N \omega'}.$$

But if one gram-molecule of substance is dissolved in 100 gram-molecules of the solvent, then

$$\theta_0 - \theta = \frac{0\cdot 02\, T_0'^2}{\omega'},$$

where T_0' is the absolute freezing point of the solvent, and ω' (which differs but slightly from ω) is the latent heat of fusion of the solvent, calculated for one gram. Now the value found for this quantity $\theta_0 - \theta$ is identical with the empirically determined quantity G. For instance, taking the solvent water, and putting T_0' as 273, and ω' as 79, the value for $\theta_0 - \theta$ is 18·8, while experiment gives $G = 18\cdot 5$.

A corresponding equation has been deduced by van't Hoff for the molecular raising of boiling point referred to 100 grams of solvent; this formula is

$$t - t_0 = \frac{0\cdot 02\, T_0^2}{w},$$

where T_0 is the absolute boiling point, and w is the latent heat of evaporation, of the solvent. This equation satisfies the results of experimental measurements.

Finally, it is to be remarked that the quantities ω and w can be calculated from the two equations, using the empirical values of G and S respectively, and that the values so calculated are the same as those obtained by direct experiments.

Now, there can be no doubt of the justness of a theory which is confirmed in so many different ways as the theory of van't Hoff is confirmed. The following statement may therefore be made. It is a natural law that *substances in solution exert the same pressure, as osmotic pressure, as that which they would manifest were they to occupy, as gases, the same volume at the same temperature;* consequently, *Avogadro's law may be applied to substances in solution.*

CHAPTER V.

AQUEOUS SOLUTIONS OF ELECTROLYTES.

ATTENTION has already been drawn to the fact that the investigations which led Raoult to the laws that have been stated in the last chapter, which laws were placed on a theoretical foundation by van't Hoff, were carried out with solutions of indifferent organic compounds in water or in organic solvents. Many substances, namely the salts, acids, and bases, and especially inorganic compounds of these classes, exhibit, when in aqueous solutions, deviations from these laws, and these deviations at first militated against the general recognition of van't Hoff's theory. The values of the quantities P, $(p-p_1)/p$, $t-t_0$, and $\theta_0-\theta$, are higher in these cases than the theory requires. The first and second quantities cannot, however, be measured with sufficient accuracy to enable any kind of regularity in their deviations from the laws to be recognised with certainty. For, on the one hand, membranes of copper ferrocyanide are no longer completely semipermeable at the high pressures that these compounds exert, and, on the other hand, the differences in the vapour pressures of aqueous solutions of the compounds are too small. But further light has been thrown on the matter by investigations into the boiling and freezing points [of solutions of these classes of compounds]. The results of some investigations, made as described in Chapter III.

are gathered together in *Tables XIII.* and *XIV.*, the same symbols being used as before.

TABLE XIII.

1	2	3	4	5	6	7	8	9	10
Substance in 100 grams water.	l	$t-t_0$	$\dfrac{t-t_0}{l}$	m	$S = \dfrac{t-t_0}{l}$	$i = \dfrac{S}{5 \cdot 2}$	$i'' = 1+(s-1)\dfrac{\lambda}{\lambda_\infty}$	$a = \dfrac{i-1}{s-1}$	$a'' = \dfrac{\lambda}{\lambda_\infty}$
Sodium chloride	5·85	0·94	0·160	58·5	9·36	1·80	1·82	0·80	0·82
,, ,,	8·80	1·40	0·159		9·30	1·78		0·78	

TABLE XIV.

1	2	3	4	5	6	7	8	9	10
Substance in 100 grams water.	l	$\theta_0-\theta$	$\dfrac{\theta_0-\theta}{l}$	m	$G = m \cdot \dfrac{\theta_0-\theta}{l}$	$i = \dfrac{G}{18 \cdot 5}$	$i'' = 1+(s-1)\dfrac{\lambda}{\lambda_\infty}$	$a' = \dfrac{i'-1}{s-1}$	$a'' = \dfrac{\lambda}{\lambda_\infty}$
Sodium chloride	5·83	3·5	0·598	58·5	34·98	1·89	1·82	0·89	0·82
,, ,,	8·80	5·2	0·591		34·51	1·87		0·87	
Potassium chloride	6·00	2·7	0·442	74·58	32·97	1·78	1·86	0·78	0·86
,, ,,	10·00	4·4	0·440		32·81	1·77		0·77	
Calcium chloride $CaCl_2 \cdot 6 H_2O$	21·90	5·2	0·237	219·98	51·90	2·69	2·50	0·84	0·75

The numbers in column 3 show that the values both of $t - t_0$ and $\theta_0 - \theta$ increase for each substance as concentration increases. Moreover, the molecular depressions of the freezing point G (column 6) agree for the solutions of sodium chloride. The values for S and G, however, are greater, on the whole, than the values for S and G in *Tables VIII.* and *X.* (pp. 104, 109). What multiples of the normal values, 5·2 and 18·5, these numbers are, is discovered by dividing the numbers by 5·2 and 18·5 respectively. The quotients thus obtained (column 7) are designated by the letter i by van't Hoff. Now it has been found that the value of i for any substance increases with increasing dilution, and that the values of i are finally expressible by whole numbers 2, 3, 4 . . . ; and, moreover, it has been shown that these values are equal for substances of similar composition; for instance, i is equal to 2 for the acids H'A', the bases B' (OH'), and the salts A'B'; and i is equal to 3 for the acids H'$_2$A", the bases B" (OH')$_2$, and the salts A'$_2$B" and A"B'$_2$. The same results are obtained whether the values of i are found by the method of osmotic pressure or by that of vapour-pressure, boiling point, or freezing point. Inasmuch as the quantities $(p-p_1)/p$, $t - t_0$, and $\theta_0 - \theta$ are proportional to the values of P, van't Hoff was led to write the general equation in the form

$$PV = iRT.$$

The solutions of all these substances behave as if a greater number of molecules were present than corresponds with the concentrations. The analogy between solutions and gases suggested an explanation of these phenomena by supposing the occurrence of a dissociation of the molecules of the dissolved substances. The anomalous behaviour of certain gases, such as nitrogen peroxide and phosphorus pentachloride, which showed a greater pressure than was demanded by Avogadro's law, had been explained by the hypothesis of dissociation. *Nevertheless, the resolution to employ this expedient in the case of solutions would hardly have been come to had not the substances in question been found to be electrolytes, and had not the theory of electrolysis supported the demand for this explanation.* As a matter of fact, abnormalities in the boiling point and freezing point phenomena appear only when the solutions conduct the electric current. A solution of sodium acetate in ether, or of potassium chloride in alcohol, behaves just as normally as an aqueous solution of sugar or urea; that is, the factor $i = 1$. But as soon as these salts are dissolved in water, and so become conductors, the factor i increases, until at an appropriate dilution it becomes equal to 2. The merit of being the first to draw attention to this circumstance belongs to Arrhenius (*Zeit. für physik. Chemie*, 1, p. 631 [1887]); his theory of electrolytic dissociation found its most important support in the fact that the values for i deduced, by it, from measurements of conductivities agreed very well with van't Hoff's values. The number i evidently expresses the ratio of the number of molecules actually present in a solution to the number which would have been present had no dissociation occurred. Now, if n gram-molecules of a substance are weighed out and dissolved in water, if the dissociation-coefficient a expresses the fraction of n which undergoes dissociation, and if z is the number of parts into which a molecule of the substance separates, then there are present in the solution $n - na$ complete molecules, and zna parts of molecules. Hence

$$i = \frac{n - na + zna}{n} = 1 + (z - 1)a;$$

and as $a = \lambda/\lambda_\infty$, when λ is the molecular conductivity at finite dilution, and λ_∞ is the molecular conductivity at infinite dilution, it follows from the conductivity of the solution that

$$i = 1 + (z - 1)\frac{\lambda}{\lambda_\infty}.$$

Column 8 of *Tables XIII.* and *XIV.* contains these values under the heading i'''. They are nearly equal to the numbers in column 7, and the agreements would have been more complete had S and G been determined for more dilute solutions. The same holds good for the values of a, when these are determined, on the one hand by the help of van't Hoff's factor i, in accordance with the equation

$$a = \frac{i - 1}{z - 1},$$

and on the other hand by the equation $a = \lambda/\lambda_\infty$. (See columns 9 and 10 of *Tables XIII.* and *XIV.*) With regard to the actual signification of the quantities a, it is to be remembered that 100 times the value of a represents the number of molecules of the substance dissociated out of every 100 molecules. The number ·89 for sodium chloride, for instance, tells that in the solution under consideration 89 per cent. of the molecules are broken up. *The parts of the molecules of a binary substance must certainly be identical with the ions, and, because of the agreement between the values found for* i *by the two methods already mentioned, the parts of the molecules of all other electrolytes must also be identical with the ions.* The fact, made apparent by the results in *Tables XIII.* and *XIV.*, that the degree of dissociation decreases with concentration, a fact which must follow from the data concerning conductivities, also points in the same direction. Arrhenius has thus explained the peculiar behaviour of aqueous solutions of electrolytes in terms of the law of van't Hoff, although that behaviour seemed at first sight to be opposed to the law; and by doing this he has not only demonstrated the applicability of Avogadro's law to electrolytic solutions, but he has also established the correctness of his own theory of dissociation.

PART III.

THE OSMOTIC THEORY OF THE CURRENT OF GALVANIC CELLS.

SEVERAL propositions have been laid down in the first and second parts of this book, the justness of which has been established almost entirely by experimental evidence. Nernst based his theory of the production of the electric current in galvanic cells on these propositions. That theory, and the deductions made from it, will be considered in this part of the book.

CHAPTER I.

LIQUID CELLS.

THE fact has been known for long that electrical differences are perceived when conductors of the second class are brought into contact. It was not, however, to be supposed that the mere contact was the cause of these phenomena.

Nernst explains the occurrence of a difference of potential between solutions, of different concentrations, of the same electrolyte, by the different velocities of the ions set in motion by osmotic pressure, whereby the kations become present in excess in one solution and the anions in the other solution. Inasmuch as the ions bear electric charges with them, an accumulation of positive electricity is brought about in one solution and of negative electricity in the other; and if two indifferent electrodes are placed in the solutions and are joined by a wire, a current must be produced in the wire because of the equalisation of the two electricities. Cells of this kind are called *liquid cells*.

In a communication submitted by H. von Helmholtz to the Prussian Academy of Sciences in 1889, Nernst calculated the differences of potential of liquid cells on the supposition that has been stated (*Sitzungsber. der Kgl. preuss. Akad. d Wiss.*, 1889, p. 83). It is necessary to make a brief repetition of these theoretical discussions, as they form the basis of the modern theory of the current. To simplify matters let it be supposed that the electrolyte is composed of two monovalent ions the velocities of migration of which are u and v respectively. Further, let p_1 be the osmotic partial pressure of the kations in the more concentrated solution, and let p_2 be that of the anions in the more dilute solution. In order that the quantity of electricity of 96,500 coulombs adhering to one gram-atom of a monovalent ion may flow through the cell, $u/(u+v)$ gram-atoms of the kation must be moved with the current, and $v/(u+v)$ gram-atoms of the anion must be moved against the current. For, when $v/(u+v)$ gram-atoms of the anion are carried from one solution to the other, by the influence of the electricity led to the anode through the wire which closes the circuit, then $v/(u+v)$ gram-atoms of the kation remain in the first solution. To these there must be added $u/(u+v)$ gram-atoms of the kation from the second solution, in order that $(u+v)/(u+v) =$ one gram-atom of the kation may reach the kathode in the first solution. Similarly, $v/(u+v)$ gram-atoms of the anion must reach the anode in order that the total quantity of the anion at the anode may be $(u+v)/(u+v) =$ one gram-atom. Now, as the volume V of a gram-molecule of a gas, in changing gradually from the pressure p_1 to the lower pressure p_2 is able to do an amount of work, at the cost of heat taken up from outside, expressed by *

$$\int_{p_1}^{p_2} V dp$$

* Let it be supposed that one gram-molecule of a gas occupies a space of 8 litres at a certain temperature and under the pressure of 5 atmospheres, and that the vessel which contains the gas has the form of a prism 80 centims. long, 10 centims. wide, and 10 centims. high. Now let so much gas escape that the pressure amounts to 4 atmospheres. The quantity of gas which has passed out of the vessel occupies the space of 8 litres at the pressure of 1 atmosphere. The gas performs a certain

so, an amount of work expressed by $\int_{p_1}^{p_2} V\,dp$ must become available when the volume of a solution, V, which contains 1 gram-atom of kation passes from the osmotic pressure p_1 to the lower pressure p_2. Therefore, $u/(u+v)$ gram-atoms of kation accomplish the work

$$\frac{u}{u+v}\int_{p_1}^{p_2} V\,dp,$$

and, as $pV = RT$, that amount of work is equal to

$$\frac{u}{u+v} RT \int_{p_1}^{p_2} \frac{dp}{p},$$

or equal to

$$\frac{u}{u+v} RT \log \frac{p_1}{p_2}.$$

But the work to be done in raising $v/u+v$ gram-atoms of anion from the pressure p_2 to the pressure p_1 is

$$\frac{v}{u+v} RT \log \frac{p_1}{p_2}.$$

The total work available from the osmotic energy is therefore

$$\frac{u-v}{u+v} RT \log \frac{p_1}{p_2}.$$

Now if this work is changed entirely into the electrical energy

amount of work during the passage out of the vessel. It raises the air resting on 800 square centims. to a height of 10 centims.; that is to say, it performs an amount of work equal to $1033 \cdot 800 \cdot 10$ gram-centims. $= 103\cdot8 \cdot 8$ kilo-decims. $= 8$ litre-atmospheres. If the same quantity of gas escapes again, the remaining gas is at the pressure of 3 atmospheres, and the total work is equal to $8(5-3) = 16$ litre-atmospheres. Generally, then, when 1 gram-molecule of a gas, occupying a space of V litres at the pressure of p_1 atmospheres, falls to the pressure of p_2 atmospheres, it performs an amount of work expressed by

$$\int_{p_1}^{p_2} V\,dp \text{ litre-atmospheres.}$$

$96,500 \cdot \pi$, when π expresses the difference of potential of the liquid cell, then

$$96,500 \cdot \pi = \frac{u-v}{u+v} R T \log \frac{p_1}{p_2}.$$

In that expression $R = 2$ gram-calories, $= 2 \cdot 4 \cdot 18 \cdot 10^7$ absolute units; 1 coulomb $= 10^{-1}$, and 1 volt $= 10^8$, absolute units. Hence

$$\pi = \frac{2 \cdot 4 \cdot 18 \cdot 10^7}{96,500 \cdot 10^{-1} \cdot 10^8} \times \frac{u-v}{u+v} \times T \log \frac{p_1}{p_2} \text{ volt},$$

$$= \cdot 0000866 \frac{u-v}{u+v} T \times \log \frac{p_1}{p_2} \text{ volt},$$

$$= \cdot 0002 \frac{u-v}{u+v} T \log \frac{p_1}{p_2} \text{ volt}$$

Hence, in order that a difference of potential may exist between two solutions of different concentrations, not only must p_1 and p_2 be different, but u and v must also differ; and the current flows from the more concentrated to the less concentrated solution when $u > v$, and in the opposite direction when $u < v$. In the case of acids u is always greater than v. A liquid cell formed of a normal solution, and a $\frac{1}{1000}$ normal solution, of hydrochloric acid, would show a difference of potential, at $17°$, equal to

$$\pi = \cdot 002 \frac{\cdot 00352 - \cdot 00069}{\cdot 00352 + \cdot 00069} \cdot 290 \cdot 3 = \cdot 117 \text{ volt.}$$

If the electrolyte contains more than two ions, the valencies of which are n_1 and n_2, the equation that has been given becomes more complicated, and assumes the form

$$\pi = \cdot 0002 \frac{\frac{u}{n_1} - \frac{v}{n_2}}{u+v} T \log \frac{p_1}{p_2} \text{ volt};$$

and if $n_1 = n_2 = n$, then, for an electrolyte of two n-valent ions,

$$\pi = \frac{\cdot 0002}{n} \cdot \frac{u-v}{u+v} T \log \frac{p_1}{p_2} \text{ volt} \ldots \ldots \ldots \text{I.}$$

A perfectly general formula, which includes the case of two different electrolytes in contact with one another, has been given by Planck; but that equation need not be considered here.

The fact is, however, to be emphasised that Nernst has confirmed his theory of liquid cells, in a more extended way, by experiments. Consequently, in such a cell osmotic energy performs electrical work. Chemical energy does not come into play in this case. One might expect that the current of this cell would continue until the osmotic pressures of the two solutions were equalised. But the ions must be constantly giving up their charges to the electrodes. And, as an electrostatic equilibrium very soon results between the kations and the anions, which equilibrium cannot be upset by the electromotive force of the cell, which is generally very small, the current of the cell must very soon cease.

The experimental demonstration of liquid cells may be omitted, because of the small values of the differences of potential of these cells.

CHAPTER II.

CONCENTRATION-CELLS.

A CONCENTRATION-CELL is formed when two bars of the same metal are placed, as electrodes, in solutions of a salt of the metal of different concentrations, these solutions being in contact. Such a cell produces a current which lasts until the concentrations are equalised; and this current is produced by the kations giving up their charges, and passing into the metallic state, at the electrode in the more concentrated solution, and so charging this electrode positively, while atoms of the other electrode go into solution as kations. The ionisation of these atoms is accompanied by changes of energy, as has been set forth previously. The kations which pass into the electrolyte carry positive electric charges with them. The electrode from which they are dissolved is therefore charged negatively, for positive electricity cannot be produced without the formation of an equal quantity of negative electricity. This process becomes clear when it is compared with the purely mechanical solution of a solid body. Just as the solid takes away heat from its surroundings in passing into solution, and, so to speak, leaves cold behind it, so do the metallic atoms take away the electric charges which they require in becoming ionised, and negative charges remain in the electrode.

As in an electrolytic cell, to the electrodes of which the

current is conducted from without, so in a galvanic cell, the electrodes are called the kathode and the anode respectively, according as the kations or the anions travel to them, and *in every case the (positive) current makes its way from the kathode through the connecting wire to the anode.* By keeping this firmly in the memory, errors in the use of the conceptions of positive and negative poles are easily avoided. The positive pole of a galvanic cell is that to which the kations proceed, and the positive pole of the source of a current is to be attached to that electrode of a decomposition-cell from which the kations must travel with the current towards the other electrode. *The positive current always goes in the same direction as the kations, both in the source of the current and also in the decomposition-cell.* Inasmuch as the kations are discharged at the kathode, the name discharging-electrode is sometimes given to the kathode, and the anode is called the solution-electrode when the anions dissolve this electrode.

Three separate differences of potential come under consideration in a concentration-cell; one is that between the two solutions, and the two others appear at the surfaces between the metal and the electrolyte. Nernst has calculated the difference of potential between the metal and the electrolyte, and he has given an explanation of the origin of this difference. For this purpose he introduced the conception of *electrolytic solution-pressure* (see contribution already cited [p. 123], and also *Zeit. für physik. Chemie*, 4, p. 129 [1889]). As a liquid evaporates from its surface until the pressure of the vapour so produced is equal to the evaporation-pressure of the liquid, so, inasmuch as evaporation and solution are analogous processes, a salt must dissolve in water until the osmotic pressure of the solution is in equilibrium with the special solution-pressure of the salt in question. Similarly there is inherent in each metal, according to Nernst, a force conditioned only by the chemical nature of the metal, which tends to dissolve metallic atoms in the form of ions. This force, which is called the solution-pressure, comes into play when the metal is immersed in an electrolyte, and it is the greater the fewer the kations already present in the solution. Putting P as the solution-pressure of a metal, and p as the osmotic pressure of the kations in the solution, three cases are to be distinguished. In the first case P may be greater than p; the metal then behaves like a quantity of salt added to an unsaturated solution of that salt. Kations tend to go into solution; and, as positive electric charges are transported by the kations while an equal quantity of negative electricity remains in the metal, the electrolyte acquires a positive, and the metal a negative, potential. However much the value of P may exceed that of p, the number of kations produced by the mere immersion of the metal in the electrolyte can be but small. For the kations collect on the bounding surface between the metal and the electrolyte, by reason of the electrostatic attraction of the kations by the negatively charged metal, and so work against the solution-pressure. Nevertheless, as soon as the free electricities are conducted away through a connecting wire, the solution-pressure manifests itself again, and continues to operate in the way already set forth until p attains the same value as P.

In the second case, when $P = p$, no difference of potential comes into play.

When, finally $P < p$, the metal corresponds with a quantity of solid salt brought into a supersaturated solution of that salt. Some of the kations now give up their charges to the metal; the metal therefore becomes charged positively, and the electrolyte acquires a negative potential. The process continues but a short time, until the positively charged metal, in its turn, repels the kations that come near it.

The theory of Nernst regards the conception of the solution-pressure of a metal as analogous with that of osmotic pressure. This hypothesis would lead us to expect that the difference of potential, $\mathfrak{p}$, between a metal and the solution of one of its salts, at a definite temperature, should be dependent only on the ratio P/p.

The following formula is arrived at by a method similar to that set forth for liquid cells:—

$$\mathfrak{p} = \frac{\cdot 0002}{n} T \log. \frac{P}{p} \text{ volt} \quad \ldots \ldots \ldots \ldots \text{ II.,}$$

when n is the valency of the kation, and $\mathfrak{p}$ is supposed to be in the direction from the metal to the solution.

In the case of a concentration-cell of the form

$$M \mid MS \text{ conc.} \mid MS \text{ dil.} \mid M,$$

where S represents the anion, the total difference of potential, π, is calculated from the algebraic sum of the three single differences of potential, by means of the equation

$$\pi = \frac{\cdot 0002}{n} T \left[\log. \frac{P}{p_1} + \frac{u-v}{u+v} \log. \frac{p_1}{p_2} - \log. \frac{P}{p_2} \right] \text{ volt}$$

$$= \frac{\cdot 0002}{n} \cdot \frac{2v}{u+v} \cdot T \log. \frac{p_1}{p_2} \text{ volt} \quad \ldots \ldots \ldots \text{ III.}$$

If the condition which is assumed in calculating π, namely, that the electrolyte is completely dissociated, is not fulfilled, the general formula runs thus:—

$$\pi = -\cdot 0002 \cdot \frac{i}{n} \cdot \frac{v}{u+v} \cdot T \log. \frac{p_1}{p_2} \text{ volt} \quad \ldots \ldots \text{ IV.,}$$

where i is van't Hoff's factor. If $u=v$ the equation takes the simpler form

$$\pi = -\cdot 0001 \frac{i}{n} T \log. \frac{p_1}{p_2} \text{ volt} \quad \ldots \ldots \ldots \text{ V.}$$

If $u>v$ then π is smaller, and if $u<v$ then π is greater, than

Formula V. represents it to be. These divergences are, however, in most cases but small. The minus sign in the formula means that the current proceeds within the concentration-cell from the more dilute to the more concentrated solution, so that the electrode of the latter becomes the kathode, as that of the former the anode.

The correctness of *Formula IV.* has been proved by different experimenters in many ways. Only one of the many experiments will be given for the more complete illustration of that formula. In the cell

$$\text{Ag/AgNO}_3 \text{ 0·1 normal/AgNO}_3 \text{ 0·01 normal/Ag}$$

$n = 1$, $u = 52$, $v = 58$, and $i = 1·87$. Hence at 18°

$$\pi = -·0002 \cdot \frac{1·87}{1} \cdot \frac{58}{52+58} \cdot 291 \log. 10 = -·0574 \text{ volt.}$$

Nernst found ·055 volt; and, considering the uncertainty of the values of u and v, this result is very satisfactory.

The production of the current in this cell has been represented as taking place in the following manner. Let K and A in figure 28 be the two silver electrodes, and let $a\,b$ be the surface of separation of the two solutions of silver nitrate L and l. When the circuit is open, K and A show a positive potential, because P has a very small value (see Chapter V.), but this potential must be greater for K than for A, in accordance with *Formula II.* When the circuit is closed, a current must, therefore, flow from K through the connecting wire to A; and at the same time silver ions separate on K and give up their charges thereat. At A, on the other hand, the atoms of the electrode take up positive charges, under the influence of the NO_3-ions, which travel from the more concentrated solution, and become ions, while negative electricity flows away through the conducting wire. For every 108 parts by weight of silver which are separated on K, 108 parts by weight of silver are dissolved from A. The process continues, with the gradual falling off of the electromotive force of the cell, until the concentrations of the solutions are equalised, and, consequently, the osmotic energy is completely exhausted.

The experimental arrangement represented in figure 28 serves to detect the concentration-current by the help of a good galvanometer. The glass tube R, which is 15 centims. long

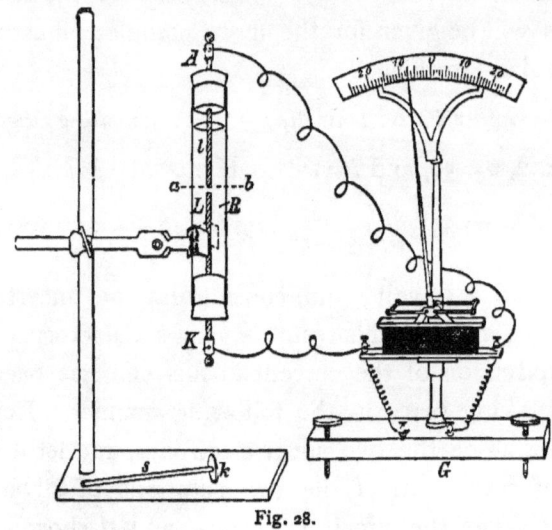

Fig. 28.

and 2·5 centims. diameter, is filled to $a\,b$ with a normal solution of silver nitrate; a layer of water is poured on to the surface of this solution by the use of the glass rod s, on the end of which is fastened the cork disc k; the electrode A is immersed in the liquid, and the lecture-galvanometer G, the magnet of which is furnished with a vertical wooden needle, is placed in the circuit.* The needle

* This galvanometer can be obtained from Keiser and Schmidt, Berlin (Johannisstrasse 20), for 55 marks. The instrument that was used in most of the following experiments, wherein weak currents had to be detected, has a resistance of 9·6 ohms, and is so delicate that a current of ·000226 ampère causes a deviation of one degree on the scale.

shows a noticeable deviation in the direction indicated by the theory. Inasmuch as very concentrated solutions of copper chloride can be prepared, the deviation shown by the cell

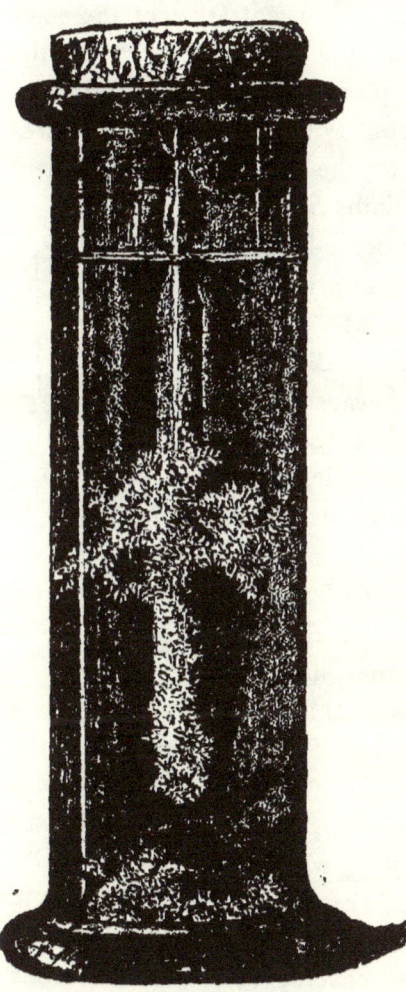

Fig. 29.

Cu/CuCl$_2$ conc./CuCl$_2$ dil./Cu

is more marked. The cell

Zn/ZnSO$_4$ conc./ZnSO$_4$ dil./Zn

is to be recommended for purposes of demonstration. In this case the electrodes assume negative potentials before the closing of the circuit, because of the great solution-pressure of the zinc. However, the potential of A is greater than that of K, and, by reason of this difference, zinc ions go into solution from A, while zinc separates on K if p_2 is sufficiently small.

Such a separation of metal in the form of a widely spreading tree is obtained, as shown in figure 29, in about a couple of hours, by filling a cylinder about 12 centims. high with a concentrated solution of stannous chloride and, after placing a layer of water on this solution, placing a long rod of tin axially in the liquid. The rod of tin represents both electrodes, and at the same time the connecting wire also. The loss of metal at the upper end of the rod is well shown in the figure.

The effect of the difference of concentration of the solutions is made very evident by the following cell:—In the H-formed tube (fig. 30) are placed the mercury electrodes A and K, which are connected with the conducting wires by platinum wires fused into the glass. The obliquely directed tube which joins the limbs S_1 and S_2 contains a stopper of glass wool. If a cold saturated solution of commercial crystallised mercurous nitrate is poured into the apparatus to the level $m\ n$, the needle of the galvanometer remains at rest, because $p_1 = p_2$. But if a concentrated solution of potassium chloride is now poured, little by little, into the limb S_1, and immediately afterwards the corresponding volume of mercurous nitrate is poured into S_2 for the purpose of restoring equilibrium, the needle shows a deviation, which is greater the greater the quantity of potassium chloride that is added. Mercurous chloride is precipitated in the limb S_1, in accordance with the equation $HgNO_3 + KCl = HgCl + KNO_3$, and the concentration of the mercury ions is hereby decreased. In this way a concentration-cell is produced, wherein it can be easily shown that, in accordance with *Formula IV.*, π increases as p_2 diminishes. In a few minutes the needle returns to the zero point, as the anode becomes covered with the mercurous chloride, which rapidly sinks to the bottom of the vessel. But if the precipitate is stirred by a glass rod, and is thus brought into suspension, the needle at once moves to its former position.

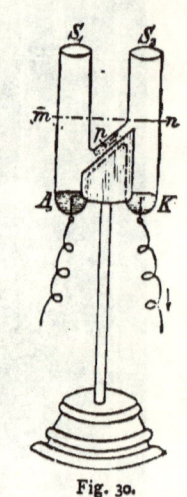

Fig. 30.

CHAPTER III.

DANIELL CELLS.

DANIELL cells
$$-M_1/M_1S/M_2S/M_2-$$
are formed from concentration-cells by making the electrodes K and A, in the experimental arrangement shown in figure 28, of the metals M_1 and M_2, and replacing the solutions L and l by solutions of the salts M_2S and M_1S, which have a common anion. The total difference of potential, π, of such a cell consists of four terms, namely

$$\pi_1 = M_1/M_1S$$
$$\pi_2 = M_1S/M_2S$$
$$\pi_3 = M_2S/M_2$$
$$\pi_4 = M_2/M_1$$

As π_2 generally amounts to not more than a few millionths of a volt, this term may be neglected in considering the total difference of potential which is, on the average, much greater; and the more nearly the concentrations of the two electrolytes and the velocities of their ions agree, the more safely may π_2 be neglected. The value of π_4 is also very small, as will be explained later. Hence it is only the potentials π_1 and π_3 that are important in calculating π. Putting P_1 and P_2 as the solution-pressures of the metals M_1 and M_2, p_1 and p_2 as the osmotic pressures relatively to the concentrations of the kations of the salts of these metals, and n the valency of the metals taken as of equal valency, it follows, in accordance with *Formula II.* (p. 131), that

$$\pi = \frac{\cdot 0002}{n} T \left(\log \frac{P_1}{p_1} - \log \frac{P_2}{p_2} \right) \text{volt} \quad \ldots \quad \ldots \quad \text{VI.}$$

$$= \frac{\cdot 0002}{n} T \left(\log \frac{P_1}{P_2} - \log \frac{p_1}{p_2} \right) \text{volt} : \quad \ldots \quad \ldots \quad \text{VII.}$$

And if the valencies of the metals are different, then

$$\pi = \cdot 0002\, T \left(\frac{1}{n_1} \log \frac{P_1}{p_1} - \frac{1}{n_2} \log \frac{P_2}{p_2} \right).$$

The direction of the current within the cell is taken as from M_1 towards M_2.

There are various methods for measuring the electromotive force of the source of a current. When the circuit is open, the measurement is made by a quadrant electrometer, by observing the deviation of the needle, proportional to the electromotive forces, which is brought about on the one hand by the cell to be tested, and on the other hand by a normal element (for instance, a Clark element $Zn/ZnSO_4/Hg_2SO_4/Hg$, for which $\pi = 1\cdot 438$ volt at $15°$). The E.M.F. of a closed cell is read off directly on a sensitive galvanometer, furnished with a high resistance (about 10^6 ohms), and graduated to volts by means of a normal element. Poggendorff's compensation-method is generally used for more accurate electrochemical work. One of the more simple forms in which this method is used is represented diagrammatically in figure 31: $a\,b$ is a measuring wire, 1 metre long, of about 50 ohms resistance; A is a constant source of current, the E.M.F. of which, E, is known and is greater than that to be measured; $b\,c$ is a variable resistance which must be sufficiently large to ensure that E does not vary when A is connected with a and c. Under these conditions the slope of the potential along the line $a\,b$ is perfectly definite; when, for instance,

$E = 2$ volts, it amounts to 0·2 volt, and hence to 0·0002 volt per millimetre of the measuring wire. The source of current B and the galvanometer G are connected with the measuring wire so that the currents meet at a. If the sliding contact d is moved until the galvanometer G indicates the absence of current, it is only necessary to multiply the number of millimetres $a\,d$ by 0·0002 to obtain the E.M.F. in volts. (For more details see Ostwald's *Physico-Chemical Measurements* [English Ed.], p. 210.)

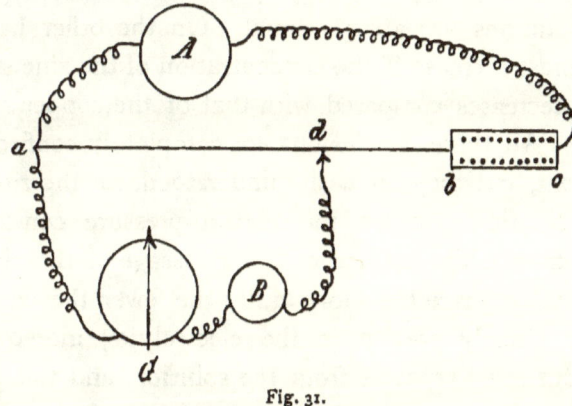

Fig. 31.

Formulæ VI. and *VII.* indicate, what is confirmed by experiment, that π is essentially independent of the nature of the anions; for instance, π is the same for the cells

$$Zn/ZnSO_4/CuSO_4/Cu$$

and

$$Zn/ZnN_2O_6/CuN_2O_6/Cu.$$

This phenomenon is made intelligible by the dissociation theory. The anion becomes important only when it very much decreases the solubility of the metallic salt. The formulæ referred to also show that the metal M_1 is the anode and M_2 is the kathode, provided the difference in brackets remains positive. This condition is fulfilled as long as the value of P_1 greatly exceeds that of P_2, and p_1 and p_2 do not differ very much from one another.

These conditions are completely fulfilled by the Daniell cell which is now commonly used,

$$Zn/ZnSO_4/CuSO_4/Cu.$$

The E.M.F. of this cell amounts to 1·114 volts at 18°, with equimolecular solutions. This value must remain unchanged, according to the theory, provided the fraction p_1/p_2 maintains the same value, that is to say, provided the concentrations of the solutions are always equal. On the other hand, the E.M.F. must increase if the concentration of the zinc sulphate solution decreases compared with that of the copper sulphate solution. All these conclusions are completely confirmed by experiment, and they are easily understood, for the zinc must go into solution because its solution-pressure considerably exceeds that of the copper, and the passage of the zinc into solution takes place the more easily the fewer the number of zinc ions already present in the electrolyte; moreover, the copper ions must separate from the solution, and this process takes place more readily from a concentrated than from a dilute solution of a salt of copper. The deviations from the normal value of 1·114 volts are never very large. For even if the concentration of one solution is a thousandfold greater than that of the other, the difference amounts only to

$$\pm \frac{0\cdot 0002}{2} \; 291 \log. 1000 = \pm 0\cdot 087 \text{ volt.}$$

An especial theoretical interest attaches to the case wherein, in a zinc-copper cell, p_2 is infinitely small. In that case $\log. \frac{P_2}{p_2}$ must be greater than $\log. \frac{P_1}{p_1}$. The sign of π is reversed, that is to say, the current passes within the cell from copper to zinc, so that the copper is the anode and the zinc the kathode. These conclusions are realised in practice when the electrolyte around the copper electrode is a solution of potassium cyanide. The electrodes K and A, of zinc and copper respectively, are fixed in the two limbs of an H-shaped tube, as shown in figure 32; the limb S_1 is filled to m with normal zinc sulphate solution (287 : 1000), and the limb S_2 is filled to n with normal potassium cyanide solution (65 : 1000); and a dilute solution of potassium sulphate, which serves as an indifferent conductor, is floated on the top of both solutions to the level $o\,p$. This cell causes a deviation of about fifteen degrees in the needle of the galvanometer G (fig. 28), in the direction already indicated. On short-circuiting, little bubbles of hydrogen come off from A. After twenty-four hours, a zinc tree has formed from the upper end of K. The changes are somewhat as follows. The copper atoms pass into the solution of potassium cyanide, which is free from Cu‥, in the form of ions, and are thus negatively charged. Cu‥Cy$_2$″ is thus produced, while the free K$_2$‥ travel to the SO$_4$″ anions of the potassium sulphate, and the K$_2$‥ of this salt to the anions SO$_4$″ of the zinc sulphate, the zinc ions of which give up their charges to the zinc electrode that serves as kathode. The Cu‥Cy$_2$″, however, reacts with 2K·Cy′, whereby the charges are mutually neutralised, in accordance with the equation

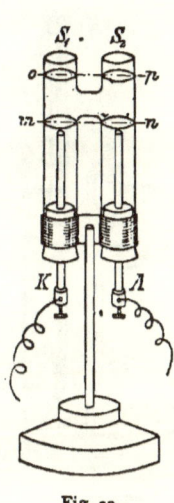

Fig. 32

$$2\text{K·Cy}' + \text{Cu‥Cy}_2'' = \text{K}_2\text{‥(CuCy}_4)''.$$

As the copper passes into the complex anion (CuCy$_4$)″, it comes about that the electrolyte around the copper electrode is free from Cu ; hence p_2 remains infinitely small.

Under the circumstances which generally obtain, log. p_1/p_2 may be put as $=0$ without introducing a serious error. *Formula VII.* (p. 137) then takes this form:

$$\pi = \frac{0\cdot 0002}{n} T \log. \frac{P_1}{P_2} \text{ volt} \quad \ldots \ldots \ldots \text{ VIII.}$$

Hence, the E.M.F. of a Daniell cell is conditioned essentially only by the ratio of the solution-pressures of the metals; and the main process that occurs in the closed cell is that *atoms of the metal with the greater solution-pressure pass, as ions, into the surrounding electrolyte, while the kations of the second electrolyte discharge themselves on the second metal.* It follows that the first metal, that is, the metal that dissolves and to which the anions of its electrolyte approach, acts as anode, and that the second metal, whereon the kations of the electrolyte thereto appertaining separate in the neutral state, acts as kathode. Inasmuch as the conception formed by Nernst of the electrolytic solution-pressures of the metals must be analogous to that of osmotic pressure, because the experiments that have so far been made verify the calculations that are based on that analogy, it follows that the impelling force of a galvanic cell, by which quantities of electricity are set in motion, must have the character of a pressure-force. In this sense Ostwald rightly speaks of a galvanic cell as a machine which is worked by osmotic pressure (or electrolytic solution-pressure).

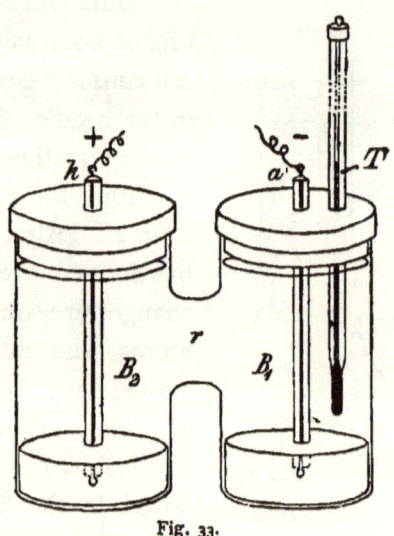

Fig. 33.

This way of looking at the matter is made clearer by the amalgam-cells of G. Meyer (*Zeit. für physik. Chemie*, **7**, p. 477 [1891]). As the amalgams are solutions of metals in mercury, it is to be supposed that the tendency of the metals to ionise their atoms will increase proportionally to the concentrations of their amalgams. Amalgams in contact with a solution of a salt of the metal in the amalgam must therefore give a current, the E.M.F. of which will be completely expressed by Equation VIII., as in this case $p_1 = p_2$. The measurements of G. Meyer agree extremely well with the theory. The mode of action of an amalgam-cell may be easily demonstrated by the cell represented in figure 33. The apparatus consists of two beakers, B_1 and B_2 (5 centims. high and 3 centims. diameter), which communicate by a tube, r, 1 centim. long and 2 centims. wide. The beakers are closed by caoutchouc covers with overlapping rims. The platinum wires a and h are fused into glass tubes and extend to the bottoms of the beakers. When 100 grams amalgam containing 1 gram zinc are placed in B_1, and 100 grams amalgam containing 0·01 gram zinc in B_2, and the vessel is filled to above r with a solution, about 10 per cent., of zinc sulphate which has been made neutral by boiling with zinc carbonate, the galvanometer included in the circuit indicates the passage of a current from the dilute to the more concentrated amalgam, as is required by the theory. The E.M.F. at 15° is

$$\pi = \frac{0·0002}{2} \cdot 288 \cdot \log \frac{1}{0·01} = 0·0576 \text{ volt.}$$

The needle, therefore, moves through 2·5 divisions of the scale. The effect of increasing the temperature may be established, very easily, by the same apparatus. It is only necessary to place the beakers in a basin filled with hot water. When the temperature of the contents of the cell has risen to 60°, as read off on the thermometer T, the deviation of the needle becomes nearly twice as great as before. In accordance with *Formula VIII.*, the E.M.F. at 60° amounts to

$$\pi = \frac{0\cdot0002}{2} \cdot 333 \cdot \log \frac{1}{0\cdot01} = 0\cdot0666 \text{ volt.}$$

Ostwald's definition of a galvanic cell as a machine driven by pressure suggests the use of an apparatus for conducting water, as represented in figure 34, for illustrating the origin and nature of the galvanic current. Although this analogy is not pertinent in every detail, nevertheless it is suitable for making clear the mutual relations of the magnitudes under consideration. The water-reservoirs A and K, and the pump P, represent the galvanic element; and the conducting tubes $a\,b\,c\,d$ represent the connecting wire. If the flow of water is stopped by a stopcock at H, and the pump is kept at rest, the water attains the same level in the tubes r_1, r_{11}, and r_{111} as in K, just as the tension in the conducting wire is equal to that at the electrode when the circuit is open. But if the stopcock is opened, the water falls gradually in the tubes, as the tension in the conducting wire of a closed cell gradually decreases. The water flows at d into the reservoir A, which may be compared to the anode, and flows with the greater force the greater is the difference of level between the vessels K and A. The mechanical energy, that is, the product of the quantity of water flowing away and the force driving it, which force is determined by the height of the fall, corresponds, if secondary conditions are neglected, with the available electrical energy. Now, if the quantity of water flowing away in unit of time, which is comparable to the quantity of current, is to remain constant, the pump P must raise an equal quantity of water in unit of time into the vessel K, and consequently must work in such a way that the difference of level, which is analogous to the difference of potential, is not changed. The difference of level is primarily dependent on the height of the stand which supports the reservoir K, so that the height of the column of water above the point of overflow may be small in proportion to the height of the stand. The manner of working of the pump corresponds to the internal resistance of the cell. If the pump works freely, a greater quantity of water can be raised into the vessel K in unit of time than if the pump works with difficulty. In the latter case the outflow stopcock must be partly closed,

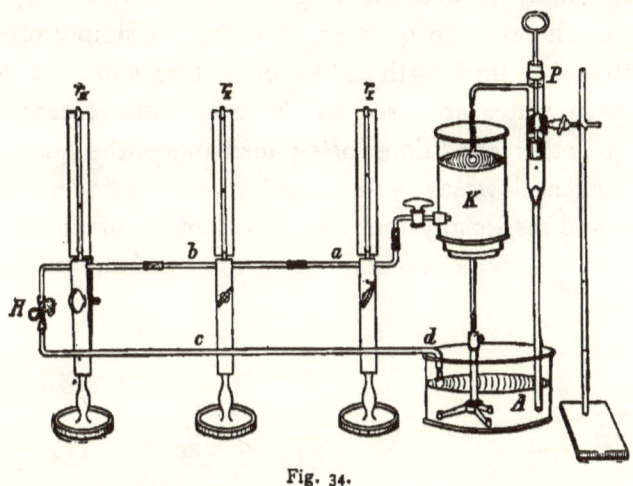

Fig. 34.

just as strong currents cannot be obtained from cells with high internal resistances (compare Daniell cells, and lead accumulators). The quantity of water which flows away depends also, as the quantity of current depends, on the resistance in the conducting apparatus. The smaller the cross-section of that apparatus the smaller will be the quantity of water that flows through it, the greater will be the energy which is lost because of the friction against the walls of the conducting tubes, and the more slowly will the pump have to work; just as a smaller amount of chemical change proceeds in an element which has a thin conducting wire. When the conducting tube has a large cross-section, the pump must work more rapidly in order to replace in K the quantity of water which flows into A; and in the analogous case, a greater quantity of chemical decomposition takes place in the substances which form a galvanic element. The resistance due to friction increases with an increase in the length of the conducting tube, and the quantity of water which flows away decreases. But the connection between the decrease in the quantity of water and the length of the tube is very complicated, whereas the quantity of current is simply inversely proportional to the length of the conducting wire. If liquids other than water are used in the apparatus represented in figure 34, other conditions being unchanged, the quantity of liquid which flows away is found to be dependent on the nature, and especially on the consistency, of the liquids, just as the specific resistances of the various metals are different.

CHAPTER IV.

REDUCTION-CELLS AND OXIDATION-CELLS.

THE explanations given in the preceding chapter might lead one to suppose that the production of the galvanic current in a Daniell cell is to be ascribed to purely physical forces. Nevertheless, as will be shown more fully in Chapter V., each metal has a special solution-pressure, which is dependent on the nature of the metal, and which must be connected in a very intimate way with the chemical affinity of the metal. Chemical energy is the ultimate source of the electrical energy which is obtained from galvanic cells. In a suitable apparatus —the galvanic cell—chemical energy is transformed into electrical energy. But as the process is essentially dependent on a solution of the metal of the anode, the result of which is the removal of kations from the electrolyte that surrounds the kathode, and as the process is then in the main an expansion of that metal, so the manner in which the chemical energy makes itself apparent in that transformation of energy is of the nature of a pressure-force.

It is certainly true that chemical processes occur in the Daniell cell. In the cell $Zn/ZnSO_4/CuSO_4/Cu$ zinc is changed into zinc sulphate, and copper is precipitated from copper sulphate. The material decomposition is expressed by saying that the zinc separates copper from the copper sulphate, while an equivalent quantity of zinc goes into solution. Whereas the chemical energy is transformed into heat when this change is caused to proceed directly by immersing a rod of zinc in a solution of copper sulphate, the chemical energy is transformed into electrical energy when the change takes place in a galvanic cell. Only in this case the process is an indirect one, in so far as it occurs in two phases at the two electrodes with the taking up and the abandonment of electrical charges ; thus

$$Zn + Zn\dot{}\dot{}SO_4\dot{}\dot{} = Zn\dot{}\dot{}SO_4\dot{}\dot{} + Zn\dot{}\dot{}$$

and

$$Zn\dot{}\dot{} + Cu\dot{}\dot{}SO_4\dot{}\dot{} = Zn\dot{}\dot{}SO_4\dot{}\dot{} + Cu.$$

The process at the anode consists in an oxidation of the zinc by the anion $SO_4\dot{}\dot{}$. The zinc itself acts as a reducing agent. The electrolyte of the kathode is reduced to copper ; it acts, therefore, while the anion $SO_4\dot{}\dot{}$ is available, as an oxidiser. The copper kathode remains chemically unchanged.

Ostwald drew especial attention to the fact that changes in the charges of the ions generally take place in all chemical processes that occur between electrolytic reducing and oxidising agents, and that these changes are a consequence of the different tendencies of the ions to take up or to give up further quantities of electricity. Moreover, he has proved that such an electromotive process can be brought about if the electrolytes are placed in separate vessels which are connected by an indifferent electrolyte, and electrodes on which these changes of charges can occur are present in the electrolyte. Platinum or carbon is a suitable material for the electrodes, because these are only required for the metallic conduction of the electricity, but do not themselves enter into any reaction with the electrolytes.

As such reduction- and oxidation-cells exhibit the transformation of chemical energy into electrical energy in a more obvious way than the Daniell cells, some of these cells may be described briefly here. Figure 35 represents a suitable experimental arrangement. The cells Z_1 and Z_2, of about 100 c.c. capacity, are easily made by cutting off the bottoms of two flasks. The necks of these cells are passed through the corks k_1 and k_2, by which they are fastened into a table-shaped stand; the electrodes K and A pass through the necks, and each is attached to a disc of platinum 3 centims. in diameter. The electrodes are put into connection with a galvanometer. A slightly acidified solution of stannous chloride (112 : 1000), which serves as the reducing agent, is poured into Z_1 till the cell is nearly filled with the liquid (an alkaline solution of stannous chloride acts better); and an acidified normal solution of common salt (58·5 : 1000) is poured into Z_2. The two cells are connected by the syphon H, which is filled with the solution of common salt. The limbs of the syphon are about 3 centims. long; they rest firmly on the nozzle-like expansions of the cells d_1 and d_2, which are barely 1 centim. apart. The needle of the galvanometer remains at rest; but as soon as a few drops of chlorine water (or better, bromine water) are placed on the platinum plate of the electrode K, by means of a pipette, the needle moves rapidly in the direction which indicates that a current is flowing from K through the external circuit to A. The $SnCl_2$ is ready to be changed to the higher state of chlorination, $SnCl_4$. For this purpose two chlorine ions are required, and two positive charges are also needed, to convert the

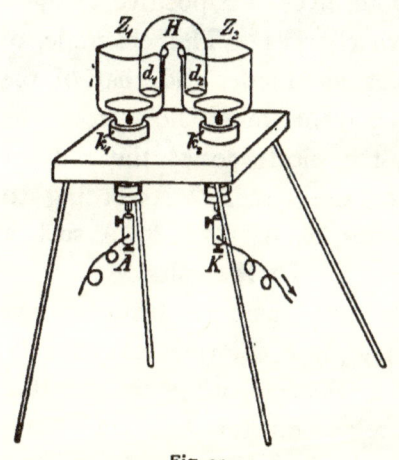

Fig. 35.

divalent $Sn^{..}$ into the tetravalent $Sn^{....}$. Two chlorine ions are produced in Z_2 from one of the chlorine molecules placed therein, and the electrode K becomes positively charged. One $Sn^{..}$ ion changes to a $Sn^{....}$ ion by taking up two positive charges. By this means A is charged negatively. The chlorine ions travel from Z_2 to Z_1. The process continues as long as molecules of chlorine are present on K. The change may be represented by the equation

$$Sn^{..}Cl_2'' + Cl_2 = Sn^{....}Cl_4'''',$$

which states that the reducer receives two positive charges, and the oxidiser two negative charges. The electrode of the former must, therefore, act as anode, and that of the latter as kathode, and the current must flow from the electrode of the oxidiser to the electrode of the reducer through the wire which closes the circuit. According to Bancroft (*Zeit. für physik. Chemie*, 10, p. 387 [1892]), such a cell as that described has an E.M.F. of 1·171 volts.

Solutions of gold chloride and mercuric chloride act similarly to chlorine water; they place chlorine ions directly at the disposal of the stannous chloride, while the metallic ions strive to discharge themselves on the kathode. The gold forms a lustrous stain on the kathode. In these cases a solution of sulphur dioxide may be used as the reducing agent, and dilute sulphuric acid as the indifferent electrolyte. The sulphur dioxide is oxidised at the cost of the oxygen of the water, the two hydrogen atoms of which are ionised in order to play the part of kations for the chlorine ions which come from Z_2. The result is that the plate A is again negatively charged. If mercuric chloride is placed on the plate K, a white precipitate of mercurous chloride, which is known to be the first phase in the reduction of mercuric chloride, is very soon seen to be forming under the plate.

The mercurous chloride cannot be produced by some sulphur dioxide which has diffused from Z_1 to Z_2, inasmuch as mercuric chloride is only reduced by a concentrated solution of sulphur dioxide, and not even then until heat is applied.

A more complicated cell, but one which is very instructive, is formed by filling Z_1 with a solution of ferrous sulphate (139 : 1000) to which 10 c.c. of sulphuric acid have been added, and Z_2 with an equimolecular solution of potassium sulphate (87 : 1000 + 10 c.c. H_2SO_4), and touching the platinum disc of the electrode K with a crystal of potassium permanganate attached by sealing-wax to the end of a glass rod. At the moment of touching the plate, the needle moves rapidly. The E.M.F. is 0·968 volt, according to Bancroft. The process is represented by the equation

$$10Fe\ddot{}SO_4'' + 2H\cdot MnO_4' + 7H_2\ddot{}SO_4'' = 5Fe_2\dddot{}(SO_4)_3\dddot{} + 2Mn\ddot{}SO_4'' + 8H_2O.$$

The O_8 of the two $HMnO_4$, by combining with the H_2 of the $2HMnO_4$ and with the $7H_2$ of the $7H_2SO_4$, forms eight molecules of water. In this way sixteen positive charges become available. Ten of these are conveyed through the connecting wire to the $10Fe\ddot{}$, which are thereby changed to ferri-ions, and the remaining six positive charges are used by the $2MnO_4'$, from which two $Mn\ddot{}$ are produced, while two positive and two negative charges are neutralised.

Various cells may be constructed with the help of the apparatus shown in figure 35, wherein electrical energy can be gained by chemical processes which are related in definite ways to those already described. In order to make the precipitation of silver chloride in accordance with the equation

$$NaCl + AgNO_3 = AgCl + NaNO_3$$

produce electrical energy, polished silver plates are laid on the platinum discs of the electrodes A and K, the cell Z_1 is filled with a solution of common salt, and the cell Z_2, and also the syphon H, with an equimolecular solution of sodium nitrate. The galvanometer fails to detect any current from the cells as thus arranged; but the needle moves as soon as a crystal of silver nitrate is placed on the silver plate of the electrode K. The result is due to the passage into solution, in the cell Z_1, of silver ions from the silver plate, and to the separation of crystals of silver on the silver plate in the cell Z_2. Now, as only a small number of silver ions can exist in the solution of common salt, the silver plate in Z_1 is very quickly covered with a film of silver chloride, which darkens in the sunlight. This cell belongs to the class of reduction- and oxidation-cells; but it may also be regarded as a concentration-cell, because the concentrations of the silver ions are very different in Z_1 and Z_2.

According to Ostwald (*Naturw. Rdsch.*, **8**, p. 573), it is possible to transform into electrical energy the chemical energy that is set free in the neutralisation of sulphuric acid and potash. A normal solution of sulphuric acid is placed in the cell Z_2, and a half-normal solution of potassium sulphate in Z_1 and also in the syphon. If a piece of palladium foil, about 4 sq. centims. area, that has been saturated, electrolytically, with hydrogen, is now laid on the platinum plate of the electrode A, and a stick of caustic potash is kept in contact with this palladium foil for a short time, little bubbles of hydrogen rise from the platinum plate of the electrode K, and the needle of the galvanometer indicates a powerful current flowing from K. The prepared palladium foil acts in this case like solid hydrogen. It is known that palladium foil has the property of occluding such large quantities of hydrogen that the hydrogen burns visibly when the foil is heated for a short time in a Bunsen burner. The hydrogen occluded by palladium, like the metals, possesses a certain striving towards ionisation, and while it becomes ionised the electrode A is negatively charged. The H· ions thus produced meet with the OH' ions of the potash and combine with them to form neutral water, while the K· ions travel to Z_2, where they compel the H· ions of the sulphuric acid to become un-ionised at the electrode K. The processes amount to the formation of water, as is also always the case in the formation of salts by the reactions of acids and bases; thus

$$2KOH + H_2SO_4 = K_2SO_4 + H_2O.$$

The peculiarity of the process in this cell is that the hydrogen which is required by the hydroxyl ions of the base for the formation of water is presented to them in the form of occluded hydrogen which must first be ionised at the anode. An extraordinary profusion of hydrogen is not required, inasmuch as the quantity of hydrogen which is used in Z_1 is obtained again in Z_2.

Ostwald classes the gas-cells, about the theory of which there has been much dispute, among reduction- and oxidation-cells, and by so doing he gives a very simple explanation of the production of the current in these cells. The hydrogen-chlorine cell is the most active. The cell Z (fig. 36), which has a capacity of about 200 c.c., is furnished with three tubuluses, t_1, t_2, and t_3. The middle tubulus must be narrow, in order that there may be but a small distance between the two side tubuluses. The tubes R_1 and R_2, which are 20 centims. long and 1 centim. wide, are placed in the tubuluses t_1 and t_2. The conducting wires from the pieces of platinised platinum foil [*]

---

[*] To platinise platinum, that is, to cover it with finely divided platinum, it is made the kathode in a decomposition-cell containing a dilute solution of platinic chloride mixed with a considerable quantity of hydrochloric acid; another piece of platinum foil is used as anode. An accumulator may be used as the source of the current.

A and K are fused through the upper ends of R_1 and R_2. The whole apparatus is filled with dilute sulphuric acid. Electrolytic hydrogen is then allowed to accumulate in R_1 to the level a, by connecting A, as kathode, and a piece of platinum foil pushed through t_3, as anode, with the source of a current. The tube R_2 is then filled to b with chlorine. When the connecting wire from K is joined to that from A, a powerful current of 1·42 volts is obtained. A single gas-element of this kind suffices to work a sensitive alarm-clock.

According to Ostwald, only those electrodes are suitable for gas-cells which are able to occlude gases. For it is only when it is occluded that a gas is in a condition to manifest its electrolytic solution-pressure, and this is especially the case with hydrogen. The current is produced in the hydrogen-chlorine cell exactly as the current is produced in the stannous chloride-chlorine cell which has been described already. Chlorine ions are readily formed in R_2 ($Cl_2 = 2Cl' + 80,200$ cals.), and by this means K becomes charged positively. These ions attract hydrogen ions from R_1, so that the occluded hydrogen in R_1 is obliged to form new ions. Hence the electrode A becomes charged negatively. While the current passes through the cell, equal volumes of hydrogen and chlorine disappear, as is required by Faraday's law. When the cell is short-circuited, the liquid rises in the tubes R_1 and R_2 about 3 centims. in an hour.

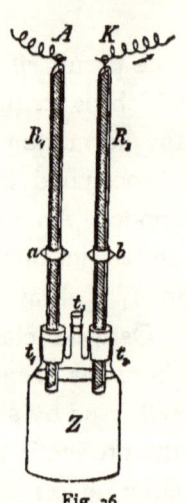

Fig. 36.

A hydrogen-oxygen cell shows the smaller E.M.F. of 1·08 volts. The explanation of the origin of the current in this cell is somewhat more complicated. Each oxygen atom on the kathode side combines with two H· ions, after these have given up their charges to K, to form water, and the available SO_4'' of the sulphuric acid causes the two atoms of occluded hydrogen of the anode tube to go into solution as ions, whereby the anode is charged negatively. Fresh quantities of hydrogen are occluded from the store of gas in proportion to the removal of the occluded hydrogen from the anode in the form of ions, and it thus comes about that the gaseous volumes of oxygen and hydrogen diminish in the proportion of 1 to 2.

Should the current of this cell not be sufficient to work an alarm-clock, which is a more suitable current-indicator than a galvanometer in a large hall, it is advisable to intercalate a relay including two dry elements, or to keep the tubulus t_3 firmly closed during the electrolytic charging of the cell. In the latter case the gases get compressed, and they give the cell a greater E.M.F. (This is the principle of the gas-accumulators, which have not, however, yet obtained practical importance.)

More powerful effects are obtained by using the following battery (fig. 37) than by employing a single gas-cell. Five narrow holes are made on the longer side, l, of a rectangular board, at a distance of 15 mm. apart, and, alternating with these, four holes are pierced on the opposite side, l'; platinum wires, which are soldered to platinum plates 43 × 47 mm. large, are brought through these holes, and the wires are fastened firmly to the two thick copper bars $a\,b$ and $c\,d$, which are fixed, by means of loops of wire, in channeled cavities cut in the board. To prevent the displacement of the parallel-arranged plates, paraffin is poured on to the underside of the board between these plates. The whole arrangement is immersed in a glass trough which is filled with dilute sulphuric acid (1 : 12). As is shown in the figure, the end a of that copper bar to which the four platinum plates are attached is connected by the commutator U (key at e) with the positive pole of a battery of two accumulators, B; and the end c of the copper bar which is attached to the five platinum plates is connected, directly, with the negative pole of the battery. The charging current produces a marked evolution of gas in the gas-battery. After charging for five minutes, the current from the gas-battery shows, at first, an E.M.F. of 1·5 volts; and after fifty minutes, and when a resistance of 1127 ohms is intercalated, the E.M.F. is still 0·7 volt. This shows that the apparatus has a relatively large capacity, which is due to the large quantity of gas that is occluded by the platinum plates. The discharging current is sufficiently powerful to

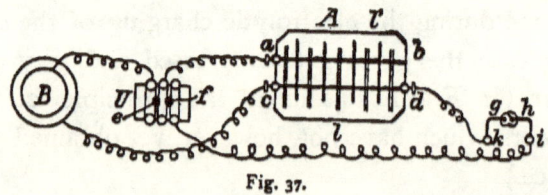

Fig. 37.

melt a thin wire of platinum 10 mm. long. This effect may be shown very well by fastening two isolated copper wires vertically on a small board, 10 mm. apart, connecting the upper ends of these wires, g and h, by a piece of thin platinum wire, surrounding the platinum with a little fresh gun-cotton, and connecting the lower ends of the copper wires i and k with d and U; if the key is put into f, after the gas-battery has been charged for one or two minutes, the gun-cotton takes fire.

In the year 1865 J. Thomsen had constructed a gas-battery of 50 cells (*Pogg. Ann.*, **124**, p. 498), each with two platinum plates, one cell of which was charged momentarily by the current from a Grove element, by an arrangement of a circular board that rotated 20 to 25 times per minute, while the other 49 cells arranged behind one another gave a constant discharging current of 49 × 1·46 = 72 volts. It is of historical interest to note that this ingenious apparatus, which was based on the principle that is applied in the accumulators of to-day, was employed in the telegraphic service at Copenhagen.

Finally, this is the place to consider the problem suggested by Ostwald (*Elektrotechnische Zeit.*, 15, p. 329 [1894]), as to how the chemical energy of carbon can be transformed directly into electrical energy. It is evident that the technical realisation of this idea, suggested by science, would be followed by consequences in comparison with which even the invention of the steam-engine would be thrown into the shade. For, on the one hand, electrical energy is that form of energy which is most easily and most completely changeable into other forms of energy; and, on the other hand, the enormous losses of energy would be avoided which are inherent in the methods now employed for producing electricity by dynamos driven by steam-power. The principle of Ostwald's "*element of the future*" has already been supplied by the gas-element. Generator-gas, such as is produced by "gasifying" carbon, would have to be supplied as the anode, and air as the kathode. Electricity would then be obtained by burning fuel, just as the chemical combination of hydrogen and oxygen is the cause of the formation of the current in the gas-element. The difficulty is to find suitable electrolytes, which shall not attack the electrodes, and which shall themselves act only as an accommodating medium without thereby being exhausted.

CHAPTER V.

THE SOLUTION-PRESSURES OF THE METALS.

ACCORDING to Nernst's theory of current-production the E.M.F. of a Daniell cell with equimolecular electrolytes is expressed by *Formula VIII.* (p. 141); if this theory is to be tested by direct trials, it is necessary to find values for the solution-pressures, P, of the metals, by experiment. There is no great difficulty in arranging the metals in a series in the order of their solution-pressures. It it only needful to determine whether one metal is able to precipitate another from the solution of one of its salts, and hence, because of a greater solution-pressure, to draw away the electrical charges to the kations of the second metal. The following is the order for some of the well-known metals: zinc, cadmium, iron, lead, copper, mercury, silver. But it has not yet been found practicable to determine the values of P quantitatively.

Nevertheless, it is possible to calculate these values by the aid of the following equation, which expresses the difference of potential $\mathfrak{p}$ between a metal and the solution of one of its salts:—

$$\mathfrak{p} = \frac{\cdot 0002}{n} T \log \frac{P}{p}.$$

The values of n, T, and p are given directly; and since an electrode has been found, in recent years, which shows no difference of potential when compared with the electrolyte, the difficulties which before stood in the way of an empirical determination of p have been overcome, at least in principle. Ostwald showed the way whereby this could be accomplished by demonstrating the behaviour of mercury dropping slowly from a capillary tube into dilute sulphuric acid.

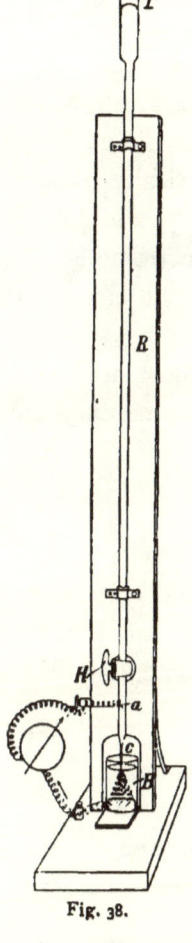

Fig. 38.

The following experiment will serve to indicate the method more fully. The glass tube R (fig. 38), which is one metre long and 8 mm. diameter, is filled with mercury. This tube is provided with a funnel, T, a stopcock, H, a platinum wire, a, fused into the glass, and a piece of thermometer tubing, c, drawn to a capillary opening about $\frac{1}{3}$ mm. wide. The beaker B contains dilute sulphuric acid (1 : 3), the surface of which is about one centimetre from c, and mercury with which a conducting wire is connected by means of a piece of platinum wire fused into the glass. When the stopcock H is opened, the stream of mercury which flows from the capillary opening divides itself into drops at one centimetre below c, and, although both electrodes are composed of the same metal, a current is detected by the galvanometer flowing from the wire which passes through the beaker.

According to the theory of Helmholtz, which is supported by the experiments of Lippmann, the current is accounted for by the relations between the electric potentials and the surface-tension of the mercury. The cell $Hg/H_2SO_4/Hg$ produces no current of itself, for the mercury electrodes show

CHAP. V.] THE SOLUTION-PRESSURES OF THE METALS. 159

equal positive potentials against the sulphuric acid which soon becomes saturated with mercurous sulphate. But when the mercury of the flowing electrode assumes the form of drops, it draws away a positive charge from the liquid, and gives this charge up again to the mercury in the beaker at the moment when the drops coalesce with this mercury.

Paschen (*Wied. Annalen*, **41**, p. 42 [1890]) has arranged the dropping electrode in such a way that the difference of potential between it and the electrolyte is practically equal to zero. By immersing the metal in the electrolyte, and joining it and the dropping electrode with an electrometer, the difference of potential between the metal and the electrolyte can be measured.

The difference of potential that has been most accurately determined, up to the present, is that between mercury and mercurous sulphate. The determination gave

$$\overrightarrow{Hg/Hg_2SO_4} = -0.99 \text{ volt.}$$

From this value, and from the value of 1·514 volts found by Wright and Thompson for the E.M.F. of the cell $Zn/ZnSO_4/Hg_2SO_4/Hg$ at 18°, the following potential difference is deduced:—

$$\overrightarrow{Zn/ZnSO_4} = 1.514 - 0.99 = +0.524 \text{ volt.}$$

This value is somewhat too large, as the zinc sulphate solution is only dissociated to 80 per cent. The value must be diminished by

$$1/2 \cdot \cdot 0002 \cdot 291 \cdot \log{^{100}/_{80}} = \cdot 003 \text{ volt,}$$

so that the corrected value $\mathfrak{p}$ is

$$\overrightarrow{Zn/ZnSO_4} = +0.521 \text{ volt.}$$

By means of the dropping electrode, Paschen obtained the number 0·5187 when he used a solution of zinc sulphate that was twice normal, and the number 0·523, which comes very near to that given above, for a normal solution of the salt.

The E.M.F., at 18°, of the cell

$$Zn/ZnSO_4 \text{ normal}/CuSO_4 \text{ normal}/Cu$$

has been determined to be 1·10 volts; hence the difference of potential for

$$\overrightarrow{CuSO_4 \text{ normal}/Cu} = 1\cdot10 - \cdot521 = +0\cdot579 \text{ volt}.$$

Hence the difference of potential for

$$\overrightarrow{Cu/CuSO_4 \text{ normal}} = -0\cdot579 \text{ volt};$$

that is to say, the electrolyte would have the potential of −·579 volt if that of the copper were equal to zero. When the degree of dissociation of the electrolyte is taken into account, the more accurate value of −·582 volt is substituted for −·579 volt; and this value has been found by experiment, namely, by determining the E.M.F. of the cell

$$Hg/Hg_2Cl_2 \text{ dissolved in } NaCl/CuSO_4 \text{ normal}/Cu.$$

The differences of potential between the various metals and normal solutions of their salts can be calculated by subtracting from ·521 the empirically determined difference of potential of the cell $Zn/ZnSO_4/MSO_4/M$.

In *Table XV.*, column *A* contains the composition of various Daniell cells arranged with equimolecular solutions, column *B* gives the E.M.F. of each of these cells as determined, at 18°, by the authors named in column *C*, and column

CHAP. V.] THE SOLUTION-PRESSURES OF THE METALS. 161

D presents the calculated and corrected differences of potential, $\mathfrak{p}$, between the metals and normal solutions of their salts.

TABLE XV.

A.	B.	C.	D.	
$\overrightarrow{\text{Zn/ZnSO}_4/\text{MgSO}_4/\text{Mg}}$	$-$ ·725 volt	Wright and Thompson	$\overrightarrow{\text{Mg/MgSO}_4}$	$= +1\cdot243$ volt
Zn/ZnSO$_4$/CdSO$_4$/Cd	$+$ ·360 ,,	F. Braun	Cd/CdSO$_4$	$= +0\cdot158$,,
Zn/ZnSO$_4$/FeSO$_4$/Fe	$+$ ·440 ,,	F. Braun	Fe/FeSO$_4$	$= +0\cdot078$,,
Zn/ZnSO$_4$/Pb(C$_2$H$_3$O$_2$)$_2$/Pb	$+$ ·607 ,,	Wright and Thompson	Pb/Pb(C$_2$H$_3$O$_2$)$_2$	$= -0\cdot089$,,
Zn/ZnSO$_4$/CuSO$_4$/Cu	$+1\cdot100$,,	F. Braun	Cu/CuSO$_4$	$= -0\cdot582$,,
Zn/ZnSO$_4$/Ag$_2$SO$_4$/Ag$_2$	$+1\cdot539$,,	Wright and Thompson	Ag$_2$/Ag$_2$SO$_4$	$= -1\cdot024$,,

The numbers in column D are characteristic magnitudes for the metals, provided that the temperature is taken as 18^c and normal solutions of the salts are employed. The nature of the anion is generally unimportant, as this need be considered only when it greatly decreases the solubility of the salt. These values may indeed be liable to errors of a few hundredths of a volt, as they are all based on a single measurement, namely, that of the difference of potential between mercury and mercurous sulphate. But the discovery of methods of measurements which shall lead to more accurate determinations of the values of $\mathfrak{p}$ is only a matter of time. The equation for $\mathfrak{p}$ indicates that the sign of these values is determined by the relation of the values of P and p; hence, as $p = 22\cdot3$ atmos. for normal solutions, the sign is dependent only on P.

The values of P can be calculated from the equation referred to. The results are presented in *Table XVI*.

11

TABLE XVI.

1	2	3	4
Metal.	Valency.	Solution-pressure, P, in atmospheres.	Weight, in grams, of metal in 1 litre of solution of the sulphate.
Mg	2	$0·115 \cdot 10^{44}$	$1·238 \cdot 10^{43}$
Zn	2	$1·786 \cdot 10^{19}$	$0·520 \cdot 10^{20}$
Cd	2	$0·599 \cdot 10^{7}$	$3·166 \cdot 10^{7}$
Fe	2	$1·068 \cdot 10^{4}$	$2·676 \cdot 10^{4}$
Pb	2	$1·950 \cdot 10^{-2}$	$1·805 \cdot 10^{-1}$
Cu	2	$2·228 \cdot 10^{-19}$	$0·313 \cdot 10^{-20}$
Hg	1	$2·178 \cdot 10^{-16}$	$0·390 \cdot 10^{-16}$
Ag	1	$0·567 \cdot 10^{-18}$	$0·547 \cdot 10^{-19}$

The values of the solution-pressures of the metals, with the exception of that of lead, are sometimes very great and sometimes very small; and, at the first glance, these values are difficult to understand. But when it is realised that the theory of Nernst represents the solution-pressure of a metal as the striving of the metal to overcome the opposing osmotic pressure of the kations which are already in solution, in order that the atoms of the metal may become ionised, then it is seen that the numbers in the third column express those osmotic pressures which the kations of the electrolyte must first possess in order that the ionisation may be impossible, and that the value of p may be equal to zero. Now this would be the case if the quantities of the metals expressed, in grams, in column 4, and calculated by means of the formula $pV = RT$, were contained in one litre of the solutions of the sulphates. But these quantities are either very large or very small, with the exception of that for lead, so that the gram-atomic weights of the metals (for mercury and silver the double atomic weights) present in one litre of the normal sulphate solutions either very much exceed, or very much fall short of, these quantities.

CHAP. V.] THE SOLUTION-PRESSURES OF THE METALS. 163

It follows, therefore, that the values of $\mathfrak{p}$ must always be positive for the metals magnesium, zinc, cadmium, and iron, and always negative for the metals copper, mercury, and silver, even although the concentrations of the electrolytes vary within the possible limits. This will be intelligible if the character of an osmotic pressure is ascribed to the solution-pressures of the metals, and the theory of electrolytic dissociation is accepted, according to which the energy consumed in separating the molecule of the electrolyte into its ions does not enter into the consideration of the question. It will be understood that the electrolyte in which zinc is immersed assumes a positive potential, and the metal itself a negative potential, because the tendency of the zinc to convey positively charged ions to the electrolyte asserts itself. The reverse of this must, however, occur with copper, for the osmotic pressure of the copper ions of the electrolyte overcomes the solution-pressure of the copper. Hence the state of affairs is such that kations give up their positive charges to the copper, and the electrolyte maintains a negative potential. Now if the cell $Zn/ZnSO_4/CuSO_4/Cu$ is closed, both the solution-pressure of the zinc and the osmotic pressure of the copper ions become permanently effective, and the result of this is that the electric current flows through the connecting wire from copper to zinc, from the kathode to the anode.

If the values of $\mathfrak{p}$ in *Table XV.* are correct, then the experimentally determined E.M.F. of a cell, π, composed of any pair of metals in that series and equimolecular solutions of their salts, must be equal to the difference of the calculated values of $\mathfrak{p}$ for these metals. Experiment confirms this conclusion. For instance, Streintz found for the cell $Cd/CdSO_4/CuSO_4/Cu$

$$\pi = 0.743 \text{ volt};$$

and the calculated value is

$$+ 0·158 + 0·579 = 0·737 \text{ volt.}$$

The observed value for the cell $Mg/MgSO_4/Ag_2SO_4/Ag$ is

$$\pi = 2·212 \text{ volts,}$$

and the calculated value is

$$\pi = + 1·243 + 1·012 = 2·255 \text{ volts.}$$

The difference between the E.M.F.s of these two cells can be easily demonstrated experimentally, by means of a galvanometer, by making the dimensions of the corresponding metals equal and using equimolecular solutions of the electrolytes. Fused rods of cadmium and magnesium, 9 mm. thick, are placed, as the anodes, in small clay cells (7 × 2 centims.). Cylindrically bent plates of copper and silver, 6 centims. high and 8 centims. wide, serve as kathodes. The electrolytes contain 10 grams cadmium sulphate, 9·3 grams copper sulphate, 9·3 grams magnesium sulphate, and 11·5 grams silver sulphate, per litre of water. The galvanometer is furnished with a resistance of 500 ohms. The cadmium-copper cell then produces a deviation of 7, and the magnesium-silver cell a deviation of 15, degrees.

A further conclusion may be drawn from this agreement between theory and practice, in accord with the theory of Nernst, that *the E.M.F. of a cell has its origin, essentially, in those places where the metals and their electrolytes come into contact with one another.*

Finally, if two different Daniell cells, which are so arranged that the kathodes of the one and the anodes of the other consist of the same metal, are disposed one behind the other, it may be expected that the total E.M.F. of the combination would be equal to the E.M.F. of a cell containing the anode of one of the cells of the combination as anode and the kathode of the

CHAP. V.] THE SOLUTION-PRESSURES OF THE METALS. 165

other as kathode. The calculated value of π for the combination of cells Zn/ZnSO$_4$/CdSO$_4$/Cd and Cd/CdSO$_4$/CuSO$_4$/Cu is

$$\pi = (\cdot 521 - \cdot 158) + (\cdot 158 + \cdot 582) = 1 \cdot 103 \text{ volts};$$

and, as a matter of fact, the same value is found for the cell Zn/ZnSO$_4$/CuSO$_4$/Cu (law of the electromotive series).

The foregoing discussion leads us to regard a list of metals arranged in order of the values of p as the true electromotive series of these metals. Such a series is essentially chemical in its character, as the solution-pressure is a chemical constant of each metal. (If the metals are arranged in conformity with their readiness to undergo oxidation, the order is the same.) The following electromotive series results from Neumann's estimations of the values of P :—

Mg, Al, Zn, Cd, Fe, Co, Ni, Pb, H, Sn, Cu, Hg, Ag, Pd, Pt, Au.

The electromotive series which were drawn up formerly should be referred to the differences of potential between the metals themselves, as these were supposed to be determined by a condenser in the air; and the opinion was held that the E.M.F. of galvanic cells was to be traced almost wholly to these differences of potential. But the fact was overlooked that the isolation of the isolating layer of the condenser could not be perfect, inasmuch as the moisture, and also the salt, in the air act as electrolytes in the experimental measurements. As a matter of fact Edlund's investigations of the Peltier's effect showed that the differences of potential arrived at by comparing the metals one with another are of small magnitude which have but slight influence on the E.M.F. of galvanic cells. The following simple experiment shows that direct contact of metals has very little effect in developing electromotive force, and that the origin of this force is much rather to be assigned to the action between metals and electrolytes. Copper wires are soldered to a nickel and a silver coin, and these wires are laid on a galvanometer; when the coins are thoroughly cleaned and then pressed one on the other, the needle remains at rest. But if a disc of blotting paper is brought between the coins, a deviation—a very small one it is true—of the needle occurs. If a drop of a solution of common salt is placed on the paper, the needle suddenly moves through about twelve divisions; after about a minute the needle certainly siwngs back to the zero point, but this is due to polarisation (see Chapter VI.).

It is therefore imperatively necessary to pay no regard to the older electromotive series; and this the more, because the production of electrical energy by the mere contact of two substances is altogether against the law of the conservation of energy. Moreover, the old contact theory gave no explanation of the part played by the electrolyte in the production of the current, nor did it indicate the significance of the chemical processes that occur in the cell.

As regards hydrogen, it is to be remarked that the general behaviour of that element shows that it must be looked on as a metal, especially as large quantities of it can be taken up by certain metals, particularly by palladium,* and the products of these actions behave like alloys. Now, if a small plate of palladium is saturated, electrolytically, with hydrogen, and is then immersed in a solution of copper sulphate, the plate is very soon covered with a lustrous layer of copper. In a similar way, gold, platinum, silver, and mercury, are precipitated; but lead, iron, cadmium, zinc, and magnesium, are not precipitated.

* Palladium is capable of occluding 936 times its own volume of hydrogen.

The following facts are in accordance with these observations: metals of the lead, iron, etc., class dissolve in acids with evolution of hydrogen, whereas metals of the class which contains gold, platinum, etc., do not evolve hydrogen from acids (overlooking such secondary reactions as those which occur with nitric acid). Hence the value of the solution-pressure of hydrogen occluded by palladium must lie between those of the solution-pressures of copper and lead. Experiments with a palladium-hydrogen plate and a norma acid solution, at 17°, showed a difference of potential equal to − ·23 volt, putting the potential of the plate as zero; hence the solution-pressure of the hydrogen is calculated to be $2·414 \cdot 10^{-3}$ atmospheres.

It is to be expected that future researches in electro-chemistry will much extend the list of the values of p. These labours, which are reserved for the future, will not only contribute to the more exact extension of the new electro-chemical theory, but, as Ostwald insists, they will also be productive of fruit in the field of pure chemistry, in so far as they will make determinations of chemical affinity possible.

The more accurate knowledge of the electromotive series of metals in electrolytes has also very great practical interest. Wherever structures made of different metals, whether alloys, or combinations of different metals in contact, or metals which are encrusted with other metals mechanically or galvanoplastically, are exposed to the influence of bodies precipitated from the atmosphere, there is a state of affairs suitable for the formation of short-circuited cells. In such cases one metal acts as the dissolving electrode and the other as the electrode which leads away the current. The former is therefore most exposed to destruction, while the latter is protected to a certain extent. An iron wire which has been covered with zinc does not rust so much at the places where the covering of zinc has been damaged, as it would do were the layer of zinc removed entirely.

A few observations on the behaviour of iron towards tin may not be out of place here, without going into any minute details. A Daniell element, made of a small clay cell, with a cylinder of sheet iron and a fused rod of tin as electrodes, and having as electrolytes equimolecular solutions of the protochlorides of the two metals, as free as possible from acid, gives a deviation of the needle of the galvanometer which remains constant for several hours, in the direction which indicates that the iron is the dissolving electrode (the anode), and, as might be expected (see Neumann's electro-chemical series, p. 165), the rod of tin becomes covered with a layer of spongy tin. But the E.M.F. of the element falls off immediately that hydrochloric acid is added to the solution of stannous chloride. If 2 volumes of hydrochloric acid of specific gravity 1·124 are added to 16 volumes of this solution, the cell gives no current, and if one volume more of the acid is added, the polarity of the cell is reversed and the tin becomes the anode.

This change of polarity may also be observed by immersing two equal-sized rods (7 mm. thick and 90 mm. long) of annealed wrought iron and pure tin, which have been polished by wet quartz sand, in solutions of acids or salts the concentrations of which range within certain limits. Figure 39, p. 172, shows the arrangement. If solutions of sulphuric or hydrochloric acid are used the strengths of which vary from $1/1000$ to $1/100$ normal (·049 to ·49 gram H_2SO_4, and ·0365 to ·365 gram HCl per litre), the iron acts as the kathode at the moment of the immersion of the rod, but it becomes the anode when the cell has been allowed to remain unclosed for ten to twenty minutes. If a normal acid solution is used, the iron acts as the electrode which conducts the current away, even when the cell is short-circuited for twenty-four hours, and tin can be detected in the solution by mercuric chloride. The change of pole is most marked in a half-normal solution of ammonium nitrate (40 : 1000), for after one minute the needle of the galvanometer moves in the opposite direction. In this case the iron acts for a moment as the kathode, and immediately afterwards as the anode. If the electrodes are dried carefully, the same phenomenon is repeated when they are again immersed in the liquid. But the iron remains as the kathode if the salt solution is considerably more dilute or more concentrated. A solution of sodium chloride produces similar results; the solution must be from normal to three times normal (58·5 to 175·5 grams per litre) if the polarity of the iron is to be reversed.

It is evident from these experiments that iron and tin come very near one another in the electromotive series. The other conditions which influence the polarity of a single cell composed of these two metals may be left undecided; it is certain that the concentrations of the electrolytes have a marked effect.

The well-known fact, that tinned iron is more liable to rust when exposed to the atmosphere than ordinary iron, is generally attributed to the occurrence of galvanic processes. If this supposition is correct, it follows that the substances precipitated from the atmosphere must act as electrolytes towards the combination of iron and tin in such a way that the iron becomes the dissolving electrode. Salts of iron must be formed, and then decomposed with the production of rust.

The following experiments may serve to confirm this view. The apparatus represented in figure 39 (p. 172) is filled with 124 c.c. of distilled water. When rods of iron and tin are immersed in the liquid and connected with the galvanometer, the needle remains at rest. Oxygen and well-washed carbon dioxide are now passed into the water; the needle remains at the zero point. But if only 1 c.c. tenth-normal sodium chloride solution (= ·00585 gram NaCl), and 1 c.c. half-normal ammonium nitrate solution (= ·04 gram NH_4NO_3) are added, the needle moves. The iron shows itself to be the anode; and when the cell has remained connected with the galvanometer for an hour, a thin yellow layer of rust is noticcable on the iron.

Again, if an iron-tin cell is filled with normal sodium chloride solution and left short-circuited for thirteen hours, flocculent masses of rust are formed containing ·003 gram iron; whereas the rust which deposits in the same time, when the iron rod alone is placed in the liquid, contains only ·0018 gram iron.

As is well known, sheet iron is tinned to protect it from rusting; and these *tin plates* are used for making all sorts of domestic utensils. But if the layer of tin is damaged and the iron is laid bare, rusting proceeds more rapidly and deeply at the exposed places than if the plates had not been tinned. If strips of tin are removed, by a knife, from a tin plate, exposure of the plate to moist air in summer for a couple of days suffices to produce red streaks of rust. But a plate of galvanised iron [that is, iron on which a coating of zinc has been deposited electrolytically] which has been treated in the same way shows no trace of rust.

CHAPTER VI.

INTENSITY OF FIXATION, AND POLARISATION.

THE results arrived at in the last chapter become of especial importance in that they afford information regarding the minimum electromotive force required for the electrolytic decomposition of a substance, and they also advance the comprehension of the phenomena of polarisation, regarding the theory of which there has been much uncertainty. When a current, whether it be such a current as is produced in a simple galvanic cell or one which flows through an electrolytic cell, causes changes in the electrolyte or on the electrodes, the current is always weakened; it is customary to refer this fact to polarisation.

Figure 39 represents an electrolytic cell filled with a normal solution of copper sulphate wherein two rods of copper, A and K, are immersed as electrodes. There is a difference of potential of ·582 volt on both electrodes, as copper ions strive to separate on both. A galvanic element may be constructed with a similar apparatus, by using a dilute solution of sulphuric acid (1 : 20) and electrodes of zinc and iron. The E.M.F. of this cell is small; but by introducing a resistance of about 1000 ohms, it is further reduced. The rod of copper A (fig. 39) is now connected with the iron-pole, and the rod K with the zinc-pole, and a galvanometer is placed in the circuit. Notwithstanding the minimum difference of potential at A and K, the current flows through the decomposition-cell, and the deviation of the needle remains constant for a long time. But the positive potential on A increases; in consequence of this, copper ions travel from A to K. The SO_4 ions cause the copper of A to send new copper ions into the electrolyte, and this in proportion to the abandonment of their charges by copper ions at K, where the positive potential is decreased by the current that is led into the cell. Hence the concentration of the electrolyte remains constant, and

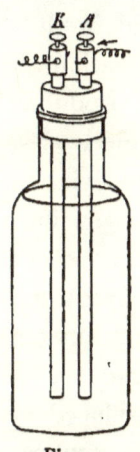

Fig. 39.

the electrodes do not suffer change. The only result of the current is to transport copper from A to K.* That a very small difference of potential suffices to do this depends on the fact that the energy which is required to form ions at the electrode A (17,500 cals. for one gram-atom of copper) is produced by the process of un-ionising at the electrode K.

An exactly similar result is produced if the copper rods A and K are replaced by rods of zinc, and a solution of zinc sulphate is used in place of the solution of copper sulphate; only the deviation of the needle is somewhat less, and, on account of the positive heat of ionisation of zinc (32,600 cals. for one gram-atom of zinc), energy is set free at A and used at K.

In neither case does the galvanometer indicate the existence of a polarisation-current when the electrolysing current is stopped, but the needle returns to the zero point. The electrodes show themselves to be *incapable of polarisation.*

* If the current that is led into the cell were stronger and continued for a longer time, a difference of concentration would soon be brought about at the electrodes, the consequence of which would be the production of a concentration-current in the direction opposite to that of the primary current.

This phenomenon is always noticeable when a primary current does not bring about any changes of composition in the decomposition-cell, and hence when the direction in which the current passes is of no importance. The same conditions obtain in the Daniell cell, and, hence, the E.M.F. of that cell is constant, provided that considerable changes in the concentration of the electrolyte do not occur in consequence of the cell being kept in use for too long a time.

But the result is different when the electrodes are insoluble, and, therefore, the anode is incapable of conveying ions into the electrolyte. Let A and K (fig. 39) be platinum electrodes, and let the electrolyte be a normal solution of sulphuric acid. If the poles of a powerful current-producer, say an accumulator-cell, are connected with these electrodes, a permanent decomposition is brought about. The deviation of the needle remains almost constant. The hydrogen ions travel to K and are set free there. The SO_4 ions betake themselves to A, and, could these ions exist in the free state, they would be un-ionised there. As a matter of fact oxygen is given off at A, in the proportion of one gram-atom of oxygen for every two gram-atoms of hydrogen produced at the kathode. The formation of oxygen is due to the neutralisation of two OH ions of the water (which is dissociated to a very small extent), in place of one SO_4 ion, as shown by the equation

$$2OH' = H_2O + O.$$

As the OH ions disappear more water is dissociated, so that the processes repeat themselves. The gases that are given off at the electrodes are partially occluded by the platinum; and when the decomposition-cell is short-circuited, these gases strive to pass again into the ionised state. In this way the state of affairs becomes that which obtains in a gas-cell, and the galvanometer indicates a polarisation-current flowing in the direction opposite to that of the primary current. The E.M.F. of the polarisation-current is not affected by the nature, or the concentration, of the oxyacid used, as these factors do not influence the occlusion of the gases.

If the weak zinc-iron cell is used to produce the primary current for the decomposition-cell $Pt/H_2SO_4/Pt$, the needle of the galvanometer shows only a slight deviation and very soon returns to the zero point. This deviation, lasting for a short time only, is due, in all probability, to the abandonment of their charges at K by a few hydrogen ions, inasmuch as the solution-pressure of hydrogen is much smaller than the osmotic pressure of the hydrogen ions that are present. But the further passage of the current is checked, because the primary current is too weak to neutralise OH ions at A, and, hence, the SO_4 ions attract the hydrogen ions from K towards A, so that the discharging of these ions is no longer possible. Still the primary current maintains a certain difference of potential at the electrodes, while the decomposition-cell behaves like a condenser. If the cell is short-circuited, a polarisation-current arises while the charges of the electrodes equalise one another, and this current causes a deviation of the needle in the opposite direction lasting for a short time.

In investigating polarisation it is very convenient, according to Friedrich C. G. Müller (*Zeit. für d. physik. und chem. Unterr.*, **8**, 1895), to bring about the reversal of the current by means of a Morse-key. The experimental arrangement will be understood from figure 40. If the primary current is taken from a single dry element, E_1, and the cell Z is composed of $Pt/H_2SO_4/Pt$, the needle of the galvanometer G_1, which indicates the direction of the primary current, very soon swings back to the zero point, and G_2 marks a polarisation-current, which lasts for a short time, in the opposite direction. The column of liquid in the manometer M does not rise, as evolution of gas does not occur. The formation of gas becomes apparent in the manometer (which is provided with a three-way cock) only when a second dry element, E_2, is intercalated, and the deviation of the needle of G_1 is then constant.

It has been observed that the primary current begins to

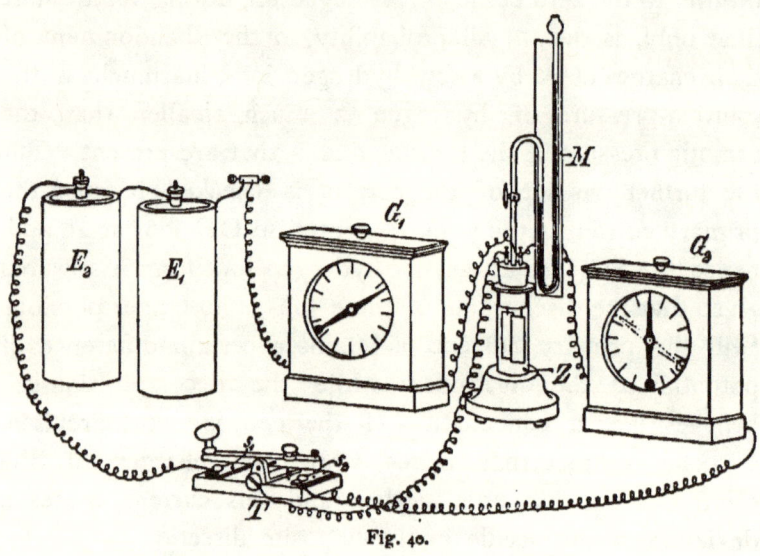

Fig. 40.

force its way permanently through the dilute sulphuric acid when the tension at A and K attains an average value of 1·6 volts.* In accordance with this it is to be supposed that a

---

* The statements about this value differ much. There are especial factors which influence this gas-polarisation: for instance, the solubilities of the gases in water, and their occlusion by the electrodes; the size, the character of the surfaces, and the distances apart, of the electrodes; the quantity of air in the electrolyte, etc. These factors have a material influence on the pressure to which the products of decomposition are exposed; and the E.M.F. of the polarisation-current increases with increasing pressure (see *Gas batteries* in Chapter IV.).

certain difference of potential is required for the neutralisation of the OH ions; this difference may be taken as

$$1{\cdot}6 + 0{\cdot}23 = 1{\cdot}83 \text{ volts,}$$

because the difference of potential between the normal sulphuric acid solution and the platinum laden with hydrogen amounts to 0·23 volt, and this aids the pressure of the primary current.

The researches of Le Blanc (*Zeit. für physik. Chemie*, **8**, 299; and **12**, 332 [1891, 1893]) show that the ions may be regarded as possessed of a definite [electromotive] force, which may be called the *intensity of fixation*, wherewith they strive to remain in the ionic condition, that is, to hold fast the electric charges corresponding with their valencies; and that a certain difference of potential, somewhat exceeding the minimum value corresponding with the intensity of fixation, must be exerted in order to overcome this force and to tear away the charges from the ions.* As regards kations, Le Blanc showed that the intensity of fixation of these is equal to the difference of potential which arises when the metal is immersed in the electrolyte. Like that difference, this intensity of fixation is dependent primarily on the solution-pressure of the metal, in accordance with the equation for $\mathfrak{p}$, but it is also influenced by the concentration of the kations and by the temperature. Hence the ions of the metals magnesium, zinc, cadmium, and iron possess a positive intensity of fixation, while the intensity of fixation of the ions of the metals lead, hydrogen, copper, mercury, and silver is negative; that is to say, the charges of the former are removed only by the expenditure of electrical energy, while the neutralisation of the latter causes electrical

---

* Le Blanc says: "By intensity of fixation I understand that electromotive force which is needed for the transference of the electric charges of ions to indifferent electrodes."—[TR.]

energy to become available. The few anions which can separate directly (Cl, Br, I) behave in a similar way.

The following table shows that the values found by Le Blanc (for normal solutions at 20°) for the polarisation of kathodes k, which values express intensities of fixation, agree well with the values of $\mathfrak{p}$ given in *Table XV*.

TABLE XVII.

	$\mathfrak{p}$	k
→Zn/Zn SO$_4$ =	+0·521	+0·515
Cd/Cd SO$_4$ =	+0·158	+0·160
Cu/Cu SO$_4$ =	−0·582	−0·560
Ag/Ag NO$_3$ =	−1·024	−1·055

f Le Blanc's theory is correct, the E.M.F. of the corresponding Daniell cells must be obtained by subtracting the two values of k.

TABLE XVIII.

Daniell element.	π calculated.	π found.
Zn/ZnSO$_4$/CuSO$_4$/Cu	1·075	1·096 (Jahn)
Zn/ZnSO$_4$/CdSO$_4$/Cd	0·355	0·360 (Braun)
Cu/CuN$_2$O$_6$/Ag$_2$N$_2$O$_6$/Ag$_2$	0·495	0·436 (Jahn)
Cd/CdSO /CuSO$_4$/Cu	0·720	0·680 (Le Blanc)

This comparison of the values of π agrees with the conclusion drawn above; hence Nernst's theory of current-production is confirmed in this way also.

The following experiment depends on the different intensities of fixation of the kations potassium and hydrogen, and is suitable for the illustration of these differences. The flask F (fig. 41) is filled to the level $m\ n$ with a dilute solution of potassium sulphate. The electrodes A and K, made of zinc and platinum respectively, are fastened through the cork k (this cork must fit firmly into the flask) in such a position that A dips about 2 centims. into the electrolyte and K is wholly immersed in the electrolyte. The cork k also carries the manometer-tube M, the stoppered funnel T, the lower end of which is drawn to a fine opening and passes to the bottom of the flask, and the piece of thin glass-rod s which is pushed into the cork after the apparatus has been arranged.

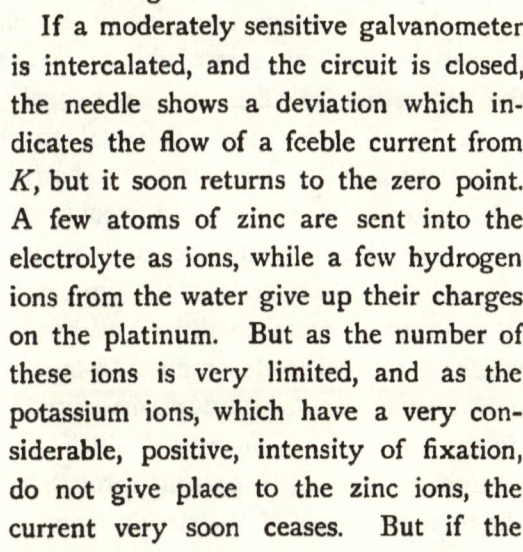

Fig. 41.

If a moderately sensitive galvanometer is intercalated, and the circuit is closed, the needle shows a deviation which indicates the flow of a feeble current from K, but it soon returns to the zero point. A few atoms of zinc are sent into the electrolyte as ions, while a few hydrogen ions from the water give up their charges on the platinum. But as the number of these ions is very limited, and as the potassium ions, which have a very considerable, positive, intensity of fixation, do not give place to the zinc ions, the current very soon ceases. But if the glass-rod s is withdrawn from the cork, and dilute sulphuric acid (coloured by indigo), of a specific gravity greater than that of the solution of potassium sulphate, is allowed to run from the stoppered funnel into the flask to about the level $o\ p$, until the surface of the acid nearly touches the rod of zinc, the needle shows a much greater deviation. At the same time little bubbles of hydrogen rise from the platinum; and if the rod s is pushed into the cork, the liquid in the manometer-tube rises slowly. There is now a sufficient number of SO_4 ions available for the potassium ions; and the hydrogen ions of the sulphuric acid, the intensity of fixation of which is small, separate in proportion to the number of zinc ions which go into solution. Under these conditions the zinc continues to dissolve, when the circuit is closed, although it is not in contact with the sulphuric acid.

The dissociation theory attributes an independent existence to the ions of the electrolyte in the electrolyte; hence, if secondary reactions are eliminated, the potential difference, π, which it is necessary to communicate to the (indifferent) electrodes for the decomposition of the electrolyte, must just exceed the algebraic sum of the intensities of fixation. Now the intensities of fixation are equal to the quantities $\mathfrak{p}_1$ and $\mathfrak{p}_2$, that is, they are equal to those differences of potential which the positive and negative constituents of the substance to be decomposed would occasion on the electrodes, at the temperature and under the conditions of concentration that obtain at the time, were those constituents to assume the ionic condition. When π reaches this value, which is called the decomposition-tension, the needle of the galvanometer is deflected permanently. The ions separate on the electrodes in quantities which are proportional to the quantity of current, i, passing through the cell; and if W is the total resistance of the circuit, then

$$i = \frac{\pi - (\mathfrak{p}_1 + \mathfrak{p}_2)}{W}.$$

Le Blanc arrived at these propositions from the results of his researches. Although the supposition that all ions of one kind have the same intensity of fixation may not be free from objections, nevertheless that theory has the advantage over others that it is able to explain the processes of electrolysis and polarisation. The theory may be illustrated experimentally by electrolysing a series of normal solutions of compounds which have a common anion, such as $ZnSO_4$, $CdSO_4$, H_2SO_4, and $CuSO_4$, between platinum electrodes, in the same decomposition-cell (fig. 12, p. 27), and with the same source of current as before; a galvanometer, and eventually a certain resistance, being included in the circuit. The deviations of the needle must increase as the electrolytes named above are used, beginning with zinc sulphate. The decomposition-tensions of the solutions of sulphate of zinc and sulphate of cadmium must be greater, and that of solution of sulphate of copper must be smaller, than that of dilute sulphuric acid, which is 1·6 volts. In each of the four cases OH ions will be discharged at the anode, a process which requires a potential-difference, $\mathfrak{p}_2$, of $+ 1·83$ volts. The theoretical decomposition-tensions of the three sulphates are given in column 3 of *Table XIX.*; column 4 contains the heats of formation of the sulphates, and column 5 the decomposition-tensions calculated therefrom.*

Table XIX.

1	2	3	4	5
Sulphate.	Values of $\mathfrak{p}_1$ for the kations.	Decomposition-tension, $\mathfrak{p}_1+\mathfrak{p}_2$.	Heat of formation, Q.	Decomposition-tension, $\pi = \dfrac{Q}{2 \times 23090}$ volts.
$ZnSO_4$	+ 0·521 volt	2·35 volts	[Zn, O, SO_3Aq]= 106,900	2·30 volts
$CdSO_4$	+ 0·158 ,,	1·98 ,,	[Cd, O, SO_3Aq]= 89,400	1·94 ,,
$CuSO_4$	− 0·582 ,,	1·24 ,,	[Cu, O, SO_3Aq]= 55,960	1·21 ,,

The agreement, to within some hundredths of a volt, between the numbers in columns 3 and 5, is in favour of the

* Regarding the formula in column 5 see Chapter IX.

CHAP. VI.] INTENSITY OF FIXATION, AND POLARISATION, 181

theory of Le Blanc, especially when it is considered that temperature-coefficients have been overlooked in calculating the numbers in column 5.

With regard to the salts of those metals whose kations are un-ionised with greater difficulty than the ions of hydrogen, it is to be remarked that the electrolysis of dilute solutions of these electrolytes takes place with differences of potential smaller than the theoretical decomposition-tensions of the salts. In such cases it is not the metals, but hydrogen ions of water, that are set free at the kathode. Nevertheless, if the electrodes are sufficiently small, the current is of high intensity, and the tension is equal to the decomposition-tension, the metals separate from the dilute solution; for, under these circumstances, there is soon a deficiency of hydrogen ions in the neighbourhood of the kathode. It is even possible to bring about the separation of the metals of the alkaline earths from aqueous solutions of chlorides of these metals by the employment of currents of great intensity.

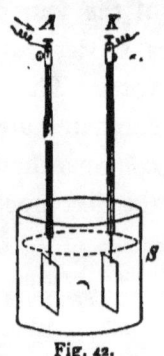

Fig. 42.

From what has been said it is evident that the kations will separate, at the kathode, from a mixture of electrolytes with a common anion in the inverse order of their intensities of fixation, when the tension at the electrodes of the electrolytic cell is gradually increased. This statement may be proved by the following experiment. A solution containing 1·24 grams copper sulphate, 110 grams ferrous sulphate, and 40 grams sulphuric acid, in 1000 grams water, is placed in the beaker S (fig. 42); the platinum electrodes A and K (5 × 4 centims.) are immersed in this liquid at a distance of 6 centims. apart, and these electrodes are connected with the poles of a battery of four accumulators, while a resistance of 40 ohms is intercalated. After a few minutes the electrode K is covered with a lustrous deposit of copper. If the electrodes are now brought nearer, and the resistance is removed, the kathode becomes covered with a velvety black metallic layer, because iron is now deposited along with copper. The presence of iron in the deposit may be shown by the formation of Prussian blue, by washing the kathode in warm hydrochloric acid, immersing it in the liquid, and afterwards adding to it some nitric acid and potassium ferrocyanide.

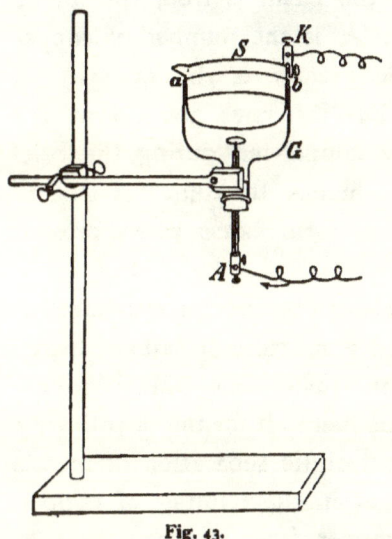

Fig. 43.

The effect on a mixture of electrolytes of increasing the tension may be demonstrated more satisfactorily by the following experiment. The vessel G (fig. 43) is a flask the bottom of which has been cut off; a small disc of platinum (1·5 centims. diameter) fastened through the neck of this flask acts as anode A; the kathode consists of a platinum basin, S, about 8 centims. diameter, placed on the rim of the vessel ab; the current is led away from this basin by the wire attached to the binding screw K. The electrolyte consists of 1000 grams water, 15·5 grams copper sulphate, 72 grams zinc sulphate, and 50 grams sulphuric acid. If the current from four accumulators is passed through a resistance of 50 to 60 ohms and then through the cell, that part of the platinum basin which is immersed in the liquid is soon covered with a lustrous layer of copper. When the resistance is removed, a dull grey spot, about 3 to 4 centims.

broad, appears opposite the anode; and if this spot is gently pressed with an agate pencil, it assumes a metallic lustre, and shows a white spot of zinc, in the centre, about 1 centim. broad, surrounded by a yellow ring of brass. Scarcely anything but zinc has separated exactly opposite the anode, because the current is stronger in the immediate neighbourhood of this position, and the small quantities of copper were already deposited. But the farther any particular part of the basin is from the anode the less is the tension, the greater is the number of copper ions still in solution, and the greater is the quantity of copper deposited thereon. Thus it comes about that the exterior ring of metal remains copper-red during the brief duration of the experiment, whereas the inner ring consists of copper and zinc which form brass when pressed together.

The knowledge that the metals can be precipitated electrolytically one after another from a mixture of salts of heavy metals has met with important applications. Metallurgical chemists now employ electrolytic methods for the quantitative estimation of metals.* The electrolytic separation of metals has become still more important in the refining of copper, and the direct extraction of copper from its ores, for that is the only method which makes it possible to obtain this metal in sufficient quantities and purity for electrotechnical purposes. The refining of the crude copper obtained by metallurgical processes is conducted in baths of copper sulphate which are kept acid by sulphuric acid. Plates are made by melting the crude copper, which may contain as much as 40 per cent. of impurities, and these plates serve as anodes; the kathodes are formed of thin sheets of pure

* For details see A. Classen's *Quantitative chemische Analyse durch Elektrolyse* [Berlin, T. Springer, 1892].

copper immersed in the baths. The work of the current, which is obtained from dynamos, is merely to carry the copper from the anodes to the kathodes. The greater part of the other substances that are contained in the crude copper is not dissolved by the anion SO_4, but falls down on the anode in the form of a mud; this mud contains silver, gold, and platinum, besides copper sulphide, lead sulphate, and basic bismuth salts. The zinc and iron which go into solution along with the copper are not precipitated on the kathode, because the tension does not exceed a certain maximum value. That the consumption of energy may be made as small as possible, the proper mixing of the baths must be attended to carefully; for loss of current is conditioned essentially only by the opposing force of the concentration-currents (overlooking electrothermic action) which would be occasioned by an accumulation of the kations at the anode.

The methods of Siemens and Höpfner are used for the direct extraction of copper from its ores. The same principle is made use of in these processes. The anode is made of copper, and the kathode of an insoluble conductor—lead or carbon; and they are separated by a diaphragm. The electrolyte is prepared by lixiviating the roasted ores. The profitableness of this method depends chiefly on the fact that salts can be used for dissolving the ores which reduce to a minimum the polarisation that always occurs with insoluble anodes, and which are brought during the electrolysis into a form wherein they are suitable for lixiviating fresh quantities of ore. Ferric sulphate, $Fe_2(SO_4)_3$, is used as a solvent in Siemens's method. This salt parts with an SO_4 group in the lixiviating vats, by which the CuS, Cu_2S, CuO, and Cu_2O in the ore are changed to sulphates; and the ferrous salt thus produced takes up an SO_4 group from the liquid surrounding the anode in the baths, in accordance with the equation

$$2FeSO_4 + SO_4 = Fe_2(SO_4)_3.$$

Höpfner employs a mixture of common salt and copper chloride solution for lixiviating the ores. Cuprous chloride is obtained in accordance with the equation

$$CuCl_2 + CuS = 2CuCl + S,$$

and this cuprous chloride remains in solution in the presence of the common salt. Copper chloride is re-formed in the neighbourhood of the anode, thus

$$2CuCl + Cl_2 = 2CuCl_2.$$

Höpfner's process has this great advantage that, because of the monovalency of copper in cuprous chloride, the quantity of metal precipitated in the baths is twice as great as that which is obtained by the same current in the process of Siemens. (*See* the experiments on Faraday's Law in *Chapter II., Part I.*)

CHAPTER VII.

IRREVERSIBLE CELLS.

WHEN zinc is placed in an acid, along with an electrode which dissolves very slightly or is altogether indifferent, and the circuit is closed, the greater solution-pressure of the zinc forces the hydrogen ions to discharge themselves at the indifferent electrode. The electricity is therefore led off from the electrode in question through the connecting wire. The current obtained from this cell is strongest immediately the cell is closed; but it soon decreases because of polarisation. For, on the one hand, the number of hydrogen ions of the electrolyte diminishes and the number of zinc ions increases; on the other hand, the internal resistance of the cell becomes greater, because the little bubbles of hydrogen adhere to the electrode which leads away the current; and, in the third place, an attraction comes into play between the adhering hydrogen and the anion of the electrolyte, whereby an electromotive force is produced which acts against the current.

The apparatus represented in figure 44 serves for the more detailed examination of these occurrences. A cylindrical roll of pure zinc, Z, to which the rod of zinc A is screwed on, is placed at the bottom of the vessel G (6 × 18 centims.). C is a plate of copper, pierced with holes like a sieve, and bent so as to be concave on the lower surface. The little rods of copper s_1, s_2, s_3, and s_4 are riveted to the under surface, and the copper rod K to the upper surface, of the copper plate. The vessel G is filled with dilute sulphuric acid (1 : 15), and is then firmly closed by the cork P, through which pass the rods A and K, and also the exit-tube for gas R. When A and K are connected by a wire which is also in connection with a moderately sensitive galvanometer, the deviation of the needle shows that K is the positive pole; and, whereas the liquid between Z and C remains clear, the liquid above C is turbid because of the little bubbles of hydrogen which rise only from C. The gradual falling off of the current is shown by the slow return of the needle towards the zero point, and also by the fact that the quantity of gas coming through R, and collected in a graduated cylinder, decreases in each period of five minutes.

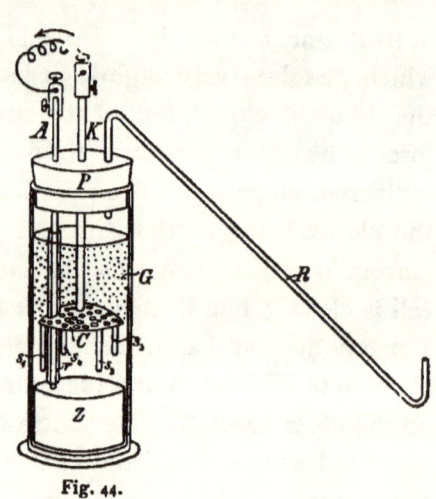

Fig. 44.

The internal resistance of the cell can be increased or diminished considerably by raising or depressing the electrode C, contact with A being prevented by the glass tube r through which A passes; and in accordance with this change of internal resistance, the deviations of the needle, and the volumes of gas which come off (5 to 15 c.c. in each five minutes), vary also. One is able in this way to demonstrate the dependence of the current strength on the resistance, as defined by Ohm's law, and also, although only approximately, the validity of Faraday's law within a source of a current. Finally, if the rod K is pressed downwards until the little rods s_1, s_2, s_3, and s_4 touch the zinc cylinder, the quantity of hydrogen reaches its maximum, and, because of the short-circuiting, the needle returns to the zero point. This phase of the experiment also illustrates the behaviour of pure and impure zinc towards acids, which has to be taken into account in the technical use of galvanic elements; for it shows that zinc, which when pure is insoluble (except at the moment of immersion) in dilute sulphuric acid at the ordinary temperature, must dissolve continuously in that acid when it comes into contact with, or is mixed with, traces of another metal.

When moderate currents are taken from cells the composition of which follows the type of the Daniell cell

$$\text{Zn}/\text{ZnSO}_4/\text{CuSO}_4/\text{Cu},$$

no other changes occur except that the concentrations of the electrolytes vary somewhat by reason of the ionisation of the zinc and the un-ionisation of the copper ions. Cells of such a kind are unpolarisable, and their E.M.F. is *constant*. Moreover, the Daniell cells are *reversible*, inasmuch as if a current equal to the primary current is sent in the opposite direction into such cells, no other action except the re-establishment of the original concentration is effected. On the other hand, cells constructed like that described above, $\text{Zn}/\text{H}_2\text{SO}_4/\text{Cu}$, behave very differently. In these cells not only the electrolyte, but also the electrode which conducts away the current, is sensibly affected, in consequence of the solution of the zinc and the disengagement of hydrogen. These cells are *inconstant and irreversible*. The original state of affairs is not restored by passing a current in the opposite direction into one of these cells. The E.M.F. of the cell can no longer be ascertained by employing the equations of Nernst. Nevertheless, the general principles of the pressure-theory can be applied to explain the production of the galvanic current from these cells.

The inconstant cells may be made constant by preventing the separation of hydrogen at the electrode which conducts away the current. This is accomplished by oxidising the hydrogen to water, after it has given up its positive charge to the kathode, by such oxidising agents as chromic acid,

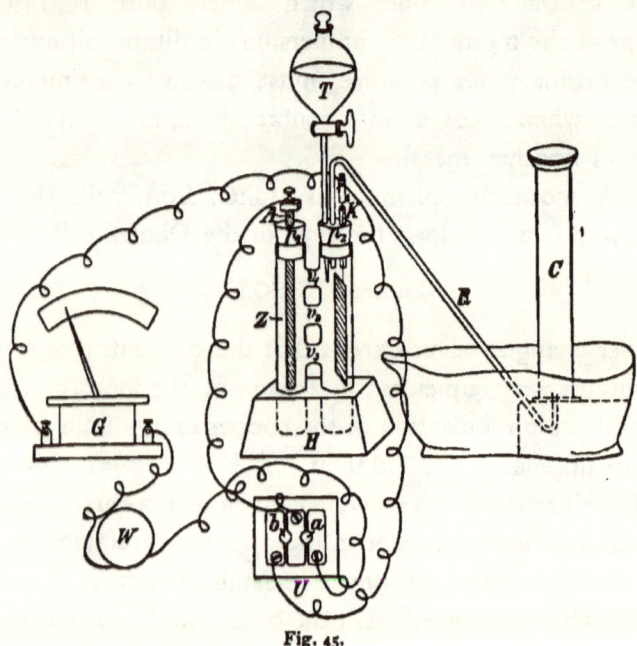

Fig. 45.

nitric acid, peroxides, etc. *Not only is the polarisation overcome by such secondary reactions, but the E.M.F. of these cells is considerably increased.* For the greater part of the chemical energy that is set free in the oxidation-process is changed into electrical energy, and the effect is the greater the more energetic is the oxidation.

These phenomena may be demonstrated in detail by the experimental arrangement represented in figure 45. The cell

Z is fastened into the wooden block H; the two limbs of the cell (2 centims. diameter) communicate by the cross-pieces v_1, v_2, and v_3. One of the limbs is closed by the cork p_1 through which passes the rod A, made of pure amalgamated zinc. The cork of the other limb, p_2, carries the platinum electrode K, the stoppered funnel T, and the tube for leading off gas R. The cell is completely filled with dilute sulphuric acid (1 : 8) by the funnel T, the narrow cylinder C is filled with water and placed over the end of the tube R, and the conducting wires are then arranged in the manner shown in the figure. Under certain circumstances it is necessary to connect a resistance coil, W, of about 500 ohms, with the galvanometer G. The galvanometer can be thrown into, or out of, the circuit by placing the key of the commutator U in a or in b. If the key is put into a, the needle shows a deviation of about seven degrees, but it soon returns some way towards zero. Small bubbles of hydrogen appear on K. If the key is now put into b, and the resistances of the galvanometer and the resistance-coil are thereby cut out, so much gas is given off that about fifteen bubbles rise per minute in C. After a few minutes the galvanometer is thrown into the circuit; the needle now moves through about five degrees only, because of polarisation. About 15 c.c. of a concentrated aqueous solution of chromium trioxide (1 : 1) are now allowed to flow from the point of the funnel T; this solution spreads through both limbs of the cell, and the evolution of hydrogen at once ceases, because the hydrogen is oxidised with formation of water and chromium oxide, as shown by the equation

$$2CrO_3 + 3H_2SO_4 + 3H_2 = Cr_2(SO_4)_3 + 6H_2O.$$

At the same time the needle very soon shows a deviation

of about twelve degrees; and the same deviation is shown after short-circuiting for an hour and a half. After the cell has been short-circuited for a couple of hours, the needle goes back to the tenth mark on the scale, because of the gradual using up of the oxidiser, and considerable quantities of chromium sulphate are now present in the electrolyte, which has become brownish black, and from which chromium hydroxide can be precipitated by adding ammonia.

Nitric acid behaves similarly to chromic acid; the nitric acid is reduced to the less oxidised compounds of nitrogen, and when its concentration has become very small, and the element begins to be polarised, it is reduced finally to ammonium nitrate.

Inasmuch as both these oxidisers are used as depolarisers in Bunsen cells, the experiment which has been described is suited for making clear the mode of action of these cells, and the processes which occur in them.

The extent to which nitric acid and sulphuric acid mixed with chromium trioxide are able to oxidise nascent hydrogen can be demonstrated more clearly by subjecting these acids to electrolysis in a U-tube (*see* fig. 12, p. 27). While the presence of oxygen can be demonstrated in the limb containing the anode, the kathode remains perfectly free from gas, and occurrence of the changes which have been indicated can be detected in the acids that surround the kathode.

The apparatus represented in figure 46 is arranged so that several oxidising agents may be brought into the electrolyte one after the other, for the purpose of bringing about depolarisation at the kathode. The vessel G is fixed in the ring of a stand; the lower end of this vessel is closed by the cork P through which passes the conducting wire A of the horizontally placed copper disc C, which is 6 centims. diameter. The small glass basin S, 3·5 centims. wide and 1 centim. high, rests on C; and the platinum disc D, the conducting wire of which, K, is held vertically by a clamp, is placed on the bottom of this basin. When the vessel G is so far filled with dilute sulphuric acid (1 : 20) that the level of the acid is half a centim. above the rim of the little basin S, the needle of a galvanometer included in the circuit shows at first a deviation of about eight degrees; but in less than two minutes it swings back to the zero point. On the other hand, as soon as small quantities of the following oxidising agents are brought, one after the other, into contact with the platinum disc, the needle shows a marked deviation, and very soon after the consumption, or the removal, of the oxidiser it returns to the zero point.

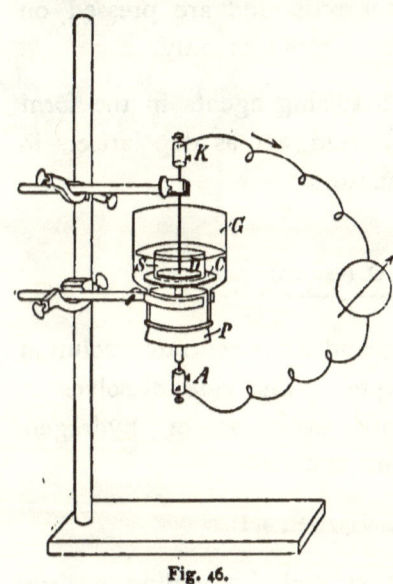

Fig. 46.

(1) A little cube, 1 c.c. in size, formed of a mixture of carbon and *pyrolusite*, containing (as shown by iodometric analysis) 6·6 per cent. available oxygen, equal to 35·7 per cent. MnO_2.*

(2) A cube of the same size cut from the peroxide plate of a Böse accumulator, containing 52 per cent. PbO_2, equal to 3·3 per cent. available oxygen.

* This material can be obtained in prisms from Keiser & Schmidt, Berlin, Johannistrasse 20.

(3) A dilute solution of gold chloride placed on the platinum plate by means of a pipette; the gold is precipitated, after a few moments, as a lustrous deposit on the platinum plate.

(4) Crystals of mercuric chloride, silver nitrate, and potassium permanganate; these crystals are fastened by sealing wax on to glass rods, and are pressed on to the platinum disc for a moment only.

It is well known that solid oxidising agents, in the form of peroxides of manganese and lead, act as depolarisers in Leclanché cells, and in accumulators.

In the Leclanché cell

$$Zn/NH_4Cl/MnO_2 \text{ (carbon)}$$

zinc is the dissolving electrode; and a concentrated solution of salammoniac is the electrolyte. The zinc dissolves in the salammoniac solution, with evolution of hydrogen, probably in accordance with the equation

$$Zn + 2NH_4Cl = ZnCl_2 \cdot 2NH_3 + H_2.$$

This process may be demonstrated by adding a little water to a mixture of ammonium chloride and zinc dust, along with a small quantity of iron powder, in a gas-evolution flask, and collecting over water the gas which comes off in considerable quantity at the ordinary temperature. In the Leclanché cell the hydrogen is oxidised by the manganese peroxide in the carbon cylinder;

$$H_2 + 2MnO_2 = H_2O + Mn_2O_3.$$

In order to recognise the depolarising effect of manganese peroxide by an experiment specially arranged for that purpose, a piece of platinum foil, A (fig. 47), and a prism

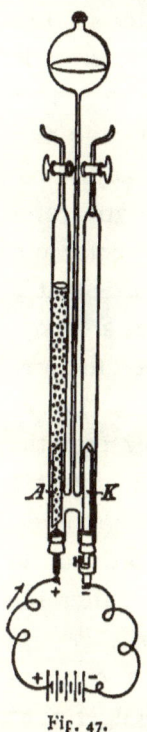

Fig. 47.

of carbon and manganese peroxide (having the same composition as the cube already described) are placed in the limbs of a Hofmann's U-tube; and the apparatus is filled with dilute sulphuric acid (1 : 12). When a current is led into the apparatus, oxygen is given off at A, but no hydrogen rises from K provided the current is weakened sufficiently by a resistance intercalated into the circuit. In this experiment the hydrogen ions are oxidised by the manganese peroxide after they have been discharged, just as happens in the Leclanché cell. If the current is made stronger, some hydrogen collects in the kathode-limb; but the volume of the hydrogen is always less than twice the volume of the oxygen which collects in the anode-limb.

The power of depolarisation of a mass of carbon and manganese peroxide is evidently the greater the stronger the current may

TABLE XX.

1	2	3	4	5
Quantity of MnO_2.	Evolution of hydrogen began with a current of	With an almost constant current-intensity of 0·072 ampères there were evolved in the first 25 minutes		
		in the voltameter	in the experimental apparatus	hence the manganese peroxide fixed
0·03 per cent.	0·056 amp.	12·66 c.c. H 6·33 c.c. O	4·59 c.c. H 5·96 c.c. O	63·74 per cent. H
2·60 per cent.	0·090 amp.	13·04 c.c. H 6·52 c.c. O	2·92 c.c. H 6·39 c.c. O	77·66 per cent. H
13·00 per cent.	0·120 amp.	12·60 c.c. H 6·30 c.c. O	0 c.c. H 5·30 c.c. O	100·00 per cent. H

be which just causes evolution of bubbles of hydrogen that is, which polarises the kathode. In order to test the capabilities of rods of carbon and manganese peroxide containing different quantities of the peroxide, a water-voltameter (of the U-form devised by Hofmann), with two platinum electrodes, a galvanometer, and an adjustable resistance, were included in the circuit of the apparatus represented in figure 47. By means of the galvanometer and the resistance, it was possible to maintain the current at a constant intensity for a long time. *Table XX.* gives a synopsis of the results of the measurements, and makes the way of working of a Leclanché element intelligible.

The numbers in columns 2, 5, 8, and 11 show that the depolarising effect of a kathode of carbon and manganese peroxide increases with increase of the quantity of peroxide, although the one increase is not directly proportional to the other, and that this effect gradually diminishes after the cell has been in use for a long time. The fact is also demonstrated that the Leclanché element gives a constant current only for a short time, but that after a longer time

TABLE XX.

6	7	8	9	10	11
With the same current of 0·072 ampères there were evolved in the following 25 minutes			In the next 75 minutes, until polarisation began, there were evolved		
in the voltameter	in the experimental apparatus	hence the MnO_2 fixed	in the voltameter	in the experimental apparatus	by a current of
12·84 c.c. H 6·42 c.c. O	4·60 c.c. H 6·19 c.c. O	61·06 per cent. H	9·40 c.c. H 4·70 c.c. O	0·50 c.c. H 4·30 c.c. O	·018 amp.
13·98 c.c. H 6·99 c.c. O	4·47 c.c. H 6·35 c.c. O	68·03 per cent. H	12·86 c.c. H 6·43 c.c. O	0·69 c.c. H 5·52 c.c. O	·024 amp.
12·90 c.c. H 6·45 c.c. O	0 c.c. H 6·36 c.c. O	100·00 per cent. H	18·38 c.c. H 9·19 c.c. O	0·92 c.c. H 8·96 c.c. O	·035 amp.

of use it is polarised the more easily the less the quantity of manganese peroxide it contains. As the cylinders of carbon and peroxide in the elements that are used in the [German] Imperial telegraphic service contain, on the average, only 1·5 per cent. MnO_2, the current that is taken from such a cell should not exceed 0·07 ampères. If a stronger current is required, several cells must be arranged parallel to one another. The opinion which is sometimes expressed that the manganese peroxide of the carbon kathode is essentially superfluous is, therefore, erroneous, inasmuch as experiment shows that pure carbon has no depolarising effect.

Columns 4, 7, and 10 of *Table XX.* show that the volume of oxygen in the experimental apparatus continues to be somewhat less than that in the voltameter. This is explained by supposing that the manganese sulphate which is produced at the kathode diffuses into the anode-limb, and is there oxidised to $Mn_2(SO_4)_3$ or to some higher oxidised compound; for, as a matter of fact, the electrolyte near the anode soon assumes a rose-red coloration.

CHAPTER VIII.

ACCUMULATORS.

THE depolarising action of lead peroxide is much more effective than that of carbon-manganese peroxide; a mass of the former compound conducts the current metallically. Let a Hofmann's apparatus for the decomposition of water, and two apparatuses constructed in the manner shown in figure 47, be included in the circuit of a battery of twenty accumulators. The kathode of one of the two apparatuses is a prism of carbon-manganese peroxide containing 35·7 per cent. MnO_2, which is equivalent to 6·6 per cent. available oxygen; the kathode of the other apparatus is a prism cut from the lead peroxide plate of a Böse accumulator containing 51·8 per cent. PbO_2, which is equivalent to 3·3 per cent. available oxygen. After twelve minutes 62 c.c. hydrogen have come off from the platinum kathode of the apparatus for decomposing water, and 41 c.c. hydrogen from the carbon-manganese peroxide kathode. But the limb that contains the kathode of lead peroxide remains quite free from gas. The reason of this behaviour is found, partly in the structures of the two oxidising agents, for the mass of lead peroxide is much less dense than that of the carbon-manganese peroxide, and there are, therefore, many more points where the hydrogen ions are attacked by the oxidiser, and partly in the different rates of reaction, for lead peroxide gives off oxygen more easily than manganese peroxide, as may be seen by heating the two substances in test-tubes. Whereas such a prism of carbon-manganese peroxide is polarised by a current of 0·063 ampères, the lead peroxide is able to sustain a current of four ampères without hydrogen being set free, that is, without the E.M.F. of the accumulator being weakened by polarisation. This is, generally, the maximum quantity of current which can be taken from an accumulator that contains a single peroxide plate, and if greater quantities of current are required, several cells must be arranged parallel to one another. The extraordinarily great depolarising action of lead peroxide explains why accumulators are so superior in steadiness even to Bunsen cells.

The foregoing experiment elucidates the processes which take place at the peroxide plate during the discharge of an accumulator, for the hydrogen is carried to that plate in the ionic state. The changes may be represented briefly by the equation

$$PbO_2 + H_2 = PbO + H_2O,$$

and the sulphuric acid then reacts on the lead oxide in accordance with the equation

$$PbO + H_2SO_4 = PbSO_4 + H_2O.$$

The processes that take place at the peroxide plate may be demonstrated by the use of a Hofmann's U-tube. In figure 48, A represents a prism cut from the lead plate of a Böse accumulator,* and K is a platinum electrode. Dilute

---

* If such a lead prism (which is certainly very brittle) is not to be had, a cylindrical rod of lead may be cast, and the surface increased by cutting longitudinal channels. Before the experiment, the outer layer must be changed to a spongy mass of lead by repeated oxidation

sulphuric acid (1 : 5) is used as the electrolyte. The lead prism is reduced, in an appropriate way, before the experiment, by leading the current from two accumulators into the apparatus for about an hour, while the stopcocks remain open (the negative pole is to be connected with A). If the accumulators are then removed, the stopcocks closed, and A and K connected with a galvanometer, a current passes from K through the connecting wire to A. Hydrogen is produced in the reaction indicated by the equation

$$Pb + H_2SO_4 = PbSO_4 + H_2;$$

the hydrogen rises from the platinum kathode, and may be recognised by the turbidity it produces in the liquid. When the galvanometer is removed, about 20 c.c. of hydrogen collect in the kathode limb after sixty minutes. If the cell is left open for some time before it is short-circuited, little bubbles of hydrogen rise slowly from A; this indicates that the porous, spongy lead is able to occlude hydrogen, and to this circumstance may be ascribed the fact that the E.M.F. of an accumulator amounts to 2·4 volts shortly after the charging, while soon afterwards it sinks to somewhat less than two volts.

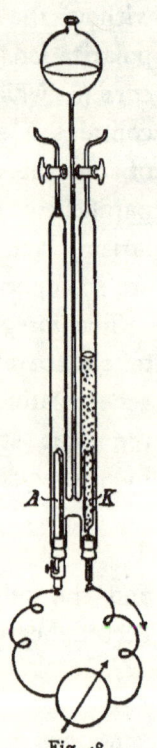

Fig. 48.

Accumulators are not only constant cells, they are also reversible cells, for they are constructed in such a way that it is possible to reverse the processes at both electrodes by leading currents into them in the reverse direction. The

and reduction, in a Hofmann's U-tube by Planté's method. The result is still better if, before this treatment, the channels are filled with a thick pulp made by mixing red lead with sulphuric acid.

layer of sulphate is reduced to lead at the lead electrode, and is oxidised to lead peroxide at the other electrode, when the accumulator is charged.

The following experiment illustrates the charging of an accumulator, and especially the production of the accumulator by the older Planté's method, the principle of which is not departed from in any essential respects in the more recent processes. Figure 49 represents a glass trough which is filled with dilute sulphuric acid (1 : 5). The leaden plates P_1, P_2, and P_3, which have been cleaned very carefully with a brush made of steel wires, are placed in the channels of the narrow sidewalls of the trough in such a way that they can be lifted out easily. As thus arranged the cell, of course, gives no current. The plates P_2 and P_3 are connected with one another; and the negative pole of a battery of at least two accumulators is connected with these plates by means of the binding screw K. The positive pole of the battery is connected with the plate P_1 by the binding screw a. When the passage of the current into the cell has been continued for about twenty minutes, both sides of the plate P_1 have become covered with a dark brown layer of peroxide; whereas the plates P_2 and P_3 appear dull grey, in consequence of the reduction of the oxide present upon them, and also because of the occlusion of hydrogen which has taken place to some extent. The current which can be obtained from the cell, even after that short period of charging, is strong enough to set an

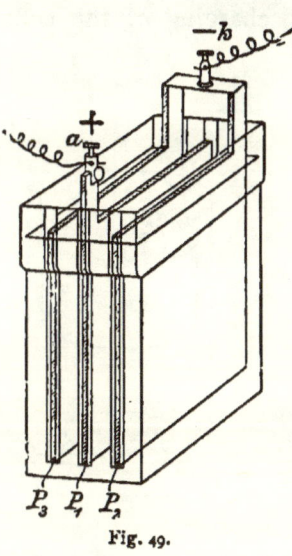

Fig. 49.

alarm-clock in action; it even suffices to decompose dilute sulphuric acid. If at least three cells, such as that shown in figure 49, are prepared, and if they are charged and discharged several times, it can be shown that the quantity of current obtained from them gradually increases. The volumes of gas, for instance, which are produced by discharging the cells through a water-voltameter, or the number of minutes which the current is able to keep alight an incandescent lamp (of about five volts), increase the more the more often the charging and discharging of the cells is performed.

If it is desired to charge a few accumulators obtained from the manufacturer for experimental use, or for some special practical purpose, and if neither the current from a dynamo nor a Gülcher's thermopile is at hand, the copper elements used in the [German] Imperial telegraphic service may be employed. These are constructed on the principle of the Meidinger cell, in the manner shown in figure 50. The

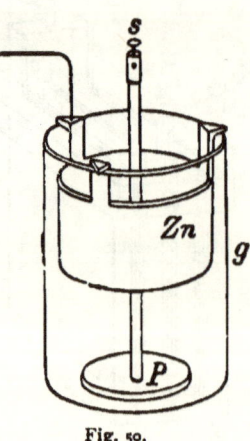

Fig. 50.

vessel g, of one litre capacity, contains a thick-walled ring of zinc, Zn, which acts as the anode, and which is suspended from the rim of the vessel by three strips of zinc. The leaden plate P, into which is cast the rod of lead s, acts as the kathode. The vessel is filled with water, in which about 20 grams of zinc sulphate are dissolved, and about 100 grams of crystals of copper sulphate are placed on the bottom of the vessel. A number of such cells is arranged in a proper way to form a battery, which is fixed in a frame and kept at rest, so that the mixing of the salt solutions may be avoided as far as possible. The E.M.F. of a copper element of this kind is one volt after a few hours; it remains constant for several weeks, and then diminishes to 0·95 volt. The internal resistance amounts to from three to eight ohms according to the concentrations of the solutions. Although the intensity of the current is small, yet it varies within very narrow limits even in the course of several months; and it is only necessary to add more or less copper sulphate to insure the proper working of the cell. The rings of zinc last for a long time. The battery of copper elements is connected, by a switch, with the accumulators that are to be charged; and after the current from the accumulators has been used, the battery is switched on again and re-charging is effected.

The following example may serve to show how the connection of such a charging battery with a battery of accumulators can be accomplished. It is required to charge eight accumulators by the use of thirty-six copper elements. Twelve copper elements are arranged in series behind one another, and the three parallel series are combined. The eight accumulators are connected in two rows of four in each, these two rows are combined parallel to one another, and the poles of the charging battery are connected in the manner shown in figure 51.

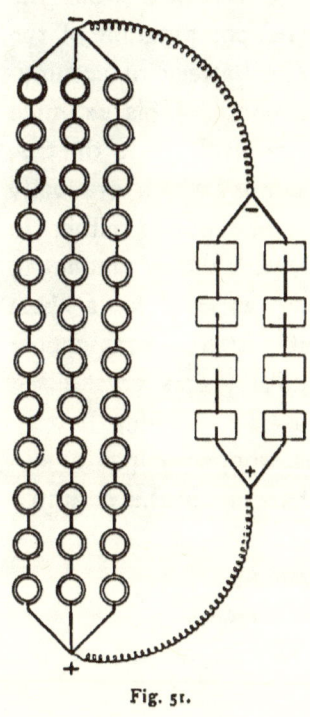

Fig. 51.

If the E.M.F. of one copper element is 1 volt, and its

internal resistance is 5 ohms, and if the opposing E.M.F. of an accumulator is 2·2 volts, then, neglecting the small internal resistance of the accumulator, the intensity of the current from the charging battery is

$$i = \frac{(12-[2\cdot2\times 4])\,36}{5\times 12^2} = \cdot 16 \text{ ampère.}$$

Now if the capacity of an accumulator amounts to 40 ampère-hours, then 80 ampère-hours must be obtained from the charging battery, because of the parallel combination of the two rows of accumulators. Hence the battery of accumulators will be charged in $80/0\cdot16 = 500$ hours. This example shows that the accumulators are storage-machines in the true sense of the word; for the electrical energy which gradually collects in them can be drawn off when required, either in small quantities or in large quantities, which may amount to 72 volt-ampères in the foregoing case, at least for a short time. Moreover, as 1 ampère separates 1·181 grams copper in one hour, and hence consumes 4·651 grams crystallised copper sulphate, and 1·213 grams zinc, the charging battery uses $3 \times 80/3 \times 12 \times \cdot004651 = 4\cdot5$ kilos. copper sulphate, and 1·2 kilos. zinc, during the process of charging, in the example given.

Let $m =$ the number of copper elements;

$e =$ the E.M.F. of one of these elements;

$w =$ the internal resistance of one of the copper elements;

$p =$ the number of copper elements combined in series (in one of the rows);

$n =$ the number of accumulator cells;

$\pi =$ the opposing E.M.F. of one of these accumulators;

$q =$ the number of accumulator cells combined in series (in one of the rows);

then, neglecting the internal resistance of the accumulator cells, the battery of accumulators will be charged by a current of the intensity

$$i = \frac{(ep - \pi q)m}{wp^2} \text{ ampères.}$$

Now let an accumulator cell have the capacity of C ampère-hours; the charging current must supply $C \cdot n/q$ ampère-hours. Putting t as the number of hours required for charging, then—

$$it = C \cdot \frac{n}{q};$$

hence—

$$t = C \cdot \frac{nw}{m} \cdot \frac{p^2}{q(ep - \pi q)} \text{ hours.}$$

From this it follows that, when C, n, w, m, e, and π remain constant, t has a minimum value when

$$p = \frac{2\pi}{e} \cdot q,$$

and, in this case,

$$t = C \cdot \frac{nw}{m} \cdot \frac{4\pi}{e^2}.$$

Hence, in order to charge as quickly as possible, the number of charging elements arranged in series ought to be about four and a half times the number of accumulator cells arranged in series in one of the rows.

It should be remarked, in regard to the zero effect of accumulators, that a distinction must be made between the quantity of current to be obtained from the accumulator, which is expressed in ampère-hours, and is dependent on the quantity of the active substance—the "active mass"—of the lead and the peroxide, and the available energy of the current to be calculated in watt-hours.* In the best type of accumulator the former amounts to 90 to 96 per cent., and the latter only to 76 to 86 per cent. A small part of the loss of energy, amounting to from 14 to 24 per cent., is due to the transformation of a certain quantity of the energy of the current into heat during the process of charging, but the greater part of the loss is dependent on the fact that the opposing E.M.F., which has to be overcome by the charging current, amounts, on the average, to 2·2 volts, whereas the mean E.M.F. of the charged accumulator is only 1·95 volts. The following example will make the meaning of these two quantities clearer. An accumulator is charged in eighteen hours with a current the average intensity of which is two ampères, and the average E.M.F. is 2·2 volts, and the accumulator is discharged in eight hours with a current of 4·25 ampères mean intensity and 1·97 volts average E.M.F. The quantity of current obtained, expressed in ampère-hours, is

$$\frac{4\cdot 25 \times 8}{2 \times 18} \times 100 = 94\cdot 4 \text{ per cent.}$$

And the energy of the current, expressed in watt-hours, is

$$\frac{1\cdot 97 \times 4\cdot 25 \times 8}{2\cdot 2 \times 2 \times 18} \times 100 = 84\cdot 5 \text{ per cent.}$$

Although accumulators are much more constant, and of greater availability, than Bunsen cells, nevertheless their efficiency falls off when they are not used for several months. The lead slowly changes to lead sulphate, with evolution of hydrogen. This self-discharge goes on more

* If the comparison of the current with the conduction of water is adhered to, the ampère-hours will correspond to the quantity of water that flows away, and the watt-hours to the product of this quantity of water into the height of the fall [*compare* pp. 143, 144].

quickly if foreign metals find their way into the accumulator, from impure sulphuric acid, for instance, or from the binding screws on the poles. The following experiment shows how far the presence of copper may cause an accumulator to become useless. Let a leaden prism (cut from the leaden plate of a Böse's accumulator) 7 centims. long, 2 centims. broad, and 0·9 centim. thick, be surrounded by platinum foil, which is isolated from the lead, and let this arrangement be immersed in dilute sulphuric acid ; then, making the leaden prism the kathode and the platinum the anode, let electrolytic reduction proceed for several hours. Now let a copper frame, R (fig. 52), to which are riveted six slips of copper foil, S, be attached by screws to the leaden prism P. This arrangement is now to be immersed in a vessel, G, filled with dilute sulphuric acid (1 : 5), and closed by a cork, K, which carries a manometer tube, M (bore = 1·3 mm.), and the little rod V. Minute bubbles of hydrogen at once begin to rise from the slips of copper. The liquid in the manometer rises about three centims. per ten minutes. The volume of gas collected in a eudiometer tube amounts to about 2·5 c.c. in three hours.*

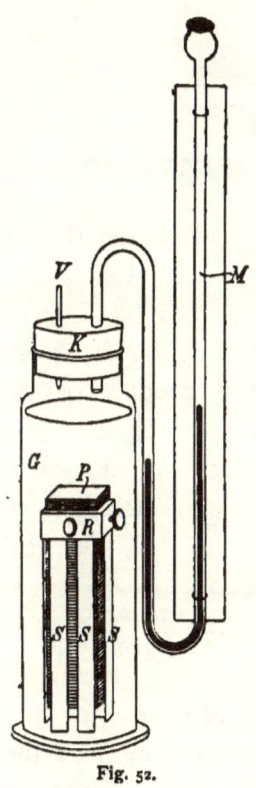

Fig. 52.

* Those who wish to become more fully acquainted with the essential features of accumulators are referred to the pamphlet, by Dr. Karl Elbs, *Die Akkumulatoren* [Leipzig: Barth ; 1893. 1 mark].

CHAPTER IX.

THE ENERGETICS OF GALVANIC ELEMENTS.

ELECTRICAL energy possesses the great advantage that it can be stored easily, and can be changed into other forms of energy with but small loss.

Just as the mechanical work which a body falling freely is able to perform is measured by the product of the impelling force, which is conditioned by the height of the fall, into the mass of the body, so electrical energy is measured by the product of the difference of potential, π, into the quantity of current, i; that is to say, electrical energy is measured in volt-coulombs. And

$$1 \text{ volt} \times 1 \text{ coulomb} = 10^7 \text{ ergs} = 0.2392 \text{ calories} = 10,314 \text{ gram-centims.}$$

When a galvanic element with the E.M.F. π, and the internal resistance w_1, is closed by a conducting wire the resistance of which is w_2, the whole of the electrical energy, $\pi \times i$, is changed into heat; and if the current circulates for t seconds with a constant E.M.F., this heat is expressed by the statement

$$C = 0.2392 \, \pi . i . t = 0.2392 \, i^2 \, (w_1 + w_2) \, t \text{ cals.}$$

The quantity of heat

$$c_1 = 0.2392 \, i^2 . w_1 . t \text{ cals.}$$

remains in the galvanic element; and the quantity of heat

$$c_2 = 0.2392\, i^2 . w_2 . t \text{ cals.}$$

is developed in the connecting wire.

Both these quantities of heat, c_1 and c_2, may be demonstrated experimentally, at any rate approximately, in a short time. A Daniell element with a porous cell is not very suitable as the source of the current, because a sufficiently powerful current is not obtained from this arrangement by reason of its relatively large internal resistance. Neither can a Bunsen's element—composed of zinc, chromic acid, and carbon—be employed, because the two electrodes produce heat directly in the electrolyte, and so heat is set free before the circuit is closed. These drawbacks are not so apparent if accumulators are employed. A Böse's accumulator, of the Y type, is suitable; it contains three plates 3·5 × 9 centims. in size, and possesses a capacity of two ampère-hours when a current of one ampère is taken from it. The double thermoscope of Looser[*] is used as the heat-indicator. This consists (*see* fig. 53) of the manometers M_1 and M_2, and the two equal-sized receivers R_1 and R_2, which are connected with the manometers by the caoutchouc tubes g_1 and g_2. Each receiver is constructed, on the principle of Bunsen's ice-calorimeter, of two cylindrical vessels set one inside the other, and having their upper rims fused together. The inner vessel, which is to contain the substance to be heated, must be 4 centims. wide and 10 centims. high, while the outer vessel has a diameter of 10 centims. The heat produced in the first vessel presses out the air that is in the jacketing space between the two vessels, so that the liquid rises in the limb of the manometer which is fastened on the scale S. [In an actual experiment] the accumulator was placed in the receiver R_1, along with 95 c.c.

[*] Obtainable from Müller & Meiswinkel, in Essen, for 45 marks.

sulphuric acid (1 : 4). The receiver R_2 contained an equal quantity of sulphuric acid; and this receiver was closed by a cork which carried two copper wires connected with a platinum wire, p, 25 centims. long and 0·1 mm. thick, and having a resistance of 2·5 ohms. When charged, the accumulator showed an E.M.F. of 1·9 volts, and an internal

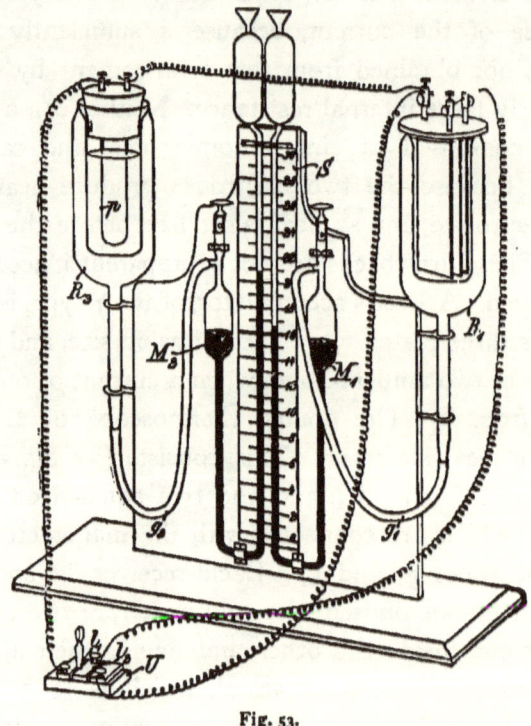

Fig. 53.

resistance of 0·15 ohm. When the liquids in the receivers had attained the temperature of the room, the wire for closing the circuit was adjusted in the way shown in figure 53 (the key of the switch U was placed in l_1). Under these conditions the quantity of current amounted to

$$i = \frac{1·9}{0·15 + 2·5} = 0·717 \text{ ampère.}$$

After twenty minutes the quantity of heat which should be developed in R_1 was

$$c_1 = 0.2392 \times .717^2 \times .15 \times 1200 = 22.13 \text{ cals.};$$

and in R_2 the quantity was

$$c_2 = 0.2392 \times .717^2 \times 2.5 \times 1200 = 368.91 \text{ cals.}$$

That is to say, the quantities of heat should have been related as $w_1 : w_2$, that is, as $1 : 16\cdot6$. As a matter of fact, the threads of liquid in the manometers rose through 4 mm. and 48 mm. respectively. Considering that heat was lost from the conducting wire, and taking into account the other, and not inconsiderable, deficiencies inherent in the process, the result was tolerably satisfactory.

When π is constant the ratio $c_2 : c_1$ increases if w_2 increases. But the absolute value of c_2 must diminish with an increment of w_2, because i decreases. In the case that π and t do not change, the value of c_2 reaches a maximum, namely

$$c_2 = 0.2392 \times i^2 \times w_2 \times t = 0.2392 \left(\frac{\pi}{w_1 + w_2}\right)^2 \times w_2 \cdot t \text{ cals.}$$

if $w_2 = w_1$; and the heat in the element is equal to the heat in the connecting wire, as is shown by experiment. The liquids in the manometers rise to equal heights, namely 12 mm. in two minutes, provided that the platinum wire p in the receiver R_2 is replaced by another wire 3 centims. long, 0·25 mm. thick, and having a resistance of 0·15 ohm. The manometers must certainly indicate a greater rise when π remains constant. Because a claim is made on the accumulator which is considerably beyond its working power, a marked polarisation sets in, and the E.M.F. falls markedly. This occurs to a yet greater degree if the accumulator is short-circuited by putting the key of the switch U into l_2. Nevertheless it is shown that almost the whole of the chemical energy of the element is changed into heat, for after short-circuiting for one minute, the liquid in the manometer M_1 has risen through 20 mm.

Moreover, the galvanic current is capable of performing work of a specific kind in apparatuses included in the circuit: the work may be mechanical, such as driving an electromotor which raises a weight; or it may be chemical, such as electrolysing a conductor of the second class, and separating therefrom constituents which, taken together, possess a greater energy-content than the compound. *In such cases the total heat of the current must be smaller, by the energy-equivalent of such work.* This also may be illustrated by experiment. A cork is placed in the receiver R_2; this cork carries a manometer fitted with a three-way cock, and two wires attached to pieces of platinum foil 1 sq. centim. area, in the manner indicated in the cell Z in figure 40 (p. 175). If these electrodes are kept 1·5 centims. apart, the resistance of the cell is 2·5 ohms, and hence as great as that of the platinum wire p in figure 53. The E.M.F. of the accumulator suffices to decompose the sulphuric acid in the receiver R_2. After a minute, the liquid in the manometer of R_2 (the bore of the tube of which is 2 mm.) shows a difference of level of 30 centims. Putting the sum of the kathodic and anodic polarisation (*compare* p. 176) as 1·6 volts, the quantity of current which flows per second through the cross-section of the path of the current is

$$\frac{1 \cdot 9 - 1 \cdot 6}{0 \cdot 15 + 2 \cdot 5} = 0 \cdot 1132 \text{ coulomb};$$

and hence, if no chemical work is done in the receiver R_2, the total heat of the current, after twenty minutes, will be

$$C = 0 \cdot 2392 \times 0 \cdot 1132 \times 1 \cdot 9 \, (0 \cdot 15 + 2 \cdot 5) \times 1200 = 61 \cdot 74 \text{ cals.}$$

But as 1 coulomb evolves 0·00009351 gram of explosive gas [2H + O], and as 68,400 cals. are required for the decomposition of 18 grams of water, the deficiency in the heat of the current brought about by the chemical work is

$$C = \frac{68400}{18} \times 0\cdot00009351 \times 0\cdot1132 \times 1200 = 48\cdot27 \text{ cals.}$$

Hence only $61\cdot74 - 48\cdot27 = 13\cdot47$ cals. are set free in the circuit. As a matter of fact, scarcely any change is to be seen in the manometers M_1 and M_2.

There is another question which concerns the relations between the electrical energy of a galvanic cell and the chemical energy from which the former takes its origin. The cell falls off in chemical energy the more the longer it produces an electric current, and the more lively are the chemical changes in the cell, that is, the greater is the thermal equivalent (expressed in calories) of the processes that occur. It was supposed for a long time that, in accordance with Thomson's rule, the electrical energy which could be gained corresponded exactly to the loss of chemical energy, in other words, that the chemical heat was equal to the current heat developed in the total circuit of the current, provided that the current was not required to perform any special work. Now if π expresses the E.M.F. of the cell, the electrical energy per gram-atom of the n-valent metal that dissolves at the anode is

$$n \times \pi \times 96,500 \times \cdot2392 = n \times \pi \times 23,090 \text{ cals.}$$

Further, if Q is the total thermal value of the chemical change referred to one gram-atom of the kation, then, according to Thomson's rule, it must be that

$$n \times \pi \times 23,090 = Q.$$

and the E.M.F. of a cell can be calculated directly from the value of Q by the equation

$$\pi = \frac{Q}{n \times 23{,}090} \text{ volts.}$$

This formula is found to be approximately accurate for the cell Zn/ZnSO$_4$/CuSO$_4$/Cu. For in the process

$$\text{Zn} + \text{CuSO}_4 = \text{ZnSO}_4 + \text{Cu}$$

$Q = 50{,}130$ cals., because [Zn, O, SO$_3$Aq] $= 106{,}090$ cals., and [Cu, O, SO$_3$Aq] $= 55{,}960$ cals. Hence it follows that

$$\pi = \frac{50{,}130}{2 \times 23{,}090} = 1\cdot09 \text{ volts,}$$

a value agreeing with the mean value obtained by actual measurements.

But considerable differences were discovered between the chemical heat and the current heat when other Daniell cells were examined accurately, so that the rule of Thomson, that is, the assertion of the complete transformability of chemical energy into electrical energy, is found to have only an approximate, and not a general, validity. More especially it is to be remarked that the E.M.F.s of the reversible cells proved to be dependent on the temperature; they increase, or decrease, more or less with change of temperature. H. von Helmholtz developed the following equation, on the basis of the second law of thermodynamics, for the connection between the absolute temperature T and the two quantities of energy, the thermal energy and the energy of the current:—

$$n \times \pi \times 23{,}090 = Q \mp 23{,}090 \times n \times T\frac{d\pi}{dT}.$$

The expression $d\pi/dT$ here denotes the temperature-coefficient of the cell, that is, the change of E.M.F. by a rise of 1° of temperature. The product $23{,}090 \times n \times T \times d\pi/dT$ is called the secondary heat, and it is equal to the difference between the current heat and the chemical heat. If $d\pi/dT$ is negative, that is, if the E.M.F. decreases with the temperature, then the electrical energy which is obtainable is less than the loss of chemical energy due to the chemical changes in the element. In this case the secondary heat is set free, besides the total heat Q, when a current is taken from the cell. On the other hand, if $d\pi/dT$ is positive, that is, if the E.M.F. increases with the temperature, the element gives a greater current energy than is equivalent to the consumption of chemical energy, and heat must be taken up from the surroundings, and converted into electrical energy, in order to keep the element at a constant temperature.

The accuracy of Helmholtz's equation has been completely verified by Jahn. A satisfactory agreement was found between the observed temperature-coefficients and those calculated from the values of π and Q which were determined by experiment.

A couple of examples will more fully elucidate Helmholtz's equation. In the cell

$$Pb/Pb(NO_3)_2/Ag_2(NO_3)_2/Ag_2$$

the following values were found for every 206·4 grams of lead dissolved at 0°:—

Current heat = 42,872 cals.
Chemical heat = 50,900 „

Hence the secondary heat is

$$23,090 \times 2 \times 273 \times d\pi/dT = -8028 \text{ cals.}$$

It follows that $d\pi/dT = -0.000636$ volt; almost the same number was observed, namely -0.000632 volt. Only 84 per cent. of the chemical energy was changed into electrical energy in this cell. The cell behaves like a steam-engine in which only a fraction of the heat set free by burning the coal is obtained as mechanical work.

The cell $Pb/Pb(C_2H_3O_2)_2/Cu(C_2H_3O_2)_2/Cu$ exemplifies the case of a positive temperature-coefficient. In this cell Jahn found

21,684 cals. for the current heat, and
17,533 „   for the chemical heat.

The value, calculated from the secondary heat, which is $+ 4151$ cals., for $d\pi/dT$ is $+ 0.000329$ volt, and the observed value was $+ 0.000385$ volt. The fifth part of the energy of the current is derived from heat outside the cell. The cell transforms into electrical energy, not only the chemical energy of the processes that take place within it, but also a considerable quantity of heat from its surroundings, and in this respect it behaves like a volume of a gas which does work in expanding at the cost of the heat of its surroundings.

Galvanic elements for which the rule of Thomson holds good must have a very small temperature-coefficient, that is to say, their E.M.F. is only very slightly influenced by the temperature. This holds not only for the element $Zn/ZnSO_4/CuSO_4/Cu$, but also for the calomel element $Zn/ZnCl_2/Hg_2Cl_2/Hg$, which is a very suitable normal element* (E.M.F. is almost exactly one volt), the temperature-coefficients of which have been determined to be $+ 0.000034$ and $+ 0.00007$ volt respectively. The E.M.F. of a reversible cell can be calculated, by Helmholtz's equation, from the thermal value of the reaction which occurs, only when the temperature-coefficient is also known. It is especially interesting to carry out the calculation for accumulators, as Streintz has determined the temperature-coefficient of these

* The Clark element, which is most commonly employed as a normal element, $Zn/ZnSO_4/Hg_2SO_4/Hg$, has an E.M.F. at 15° of 1·4336 volts; and its temperature-coefficient amounts to $-$ 0·001 volt.

cells to be 0·00032 volt, and Tschelzoff has found the heat of formation of lead peroxide [PbO, O] to be 12,140 cals. The thermal value of the processes occurring at the anode and kathode of an accumulator, referred to one gram-atom of lead, is found in the following way:—

I. At the anode.

$$SO_4 + H_2O = H_2SO_4 + O = -68,400 \text{ cals.}$$
$$Pb + O = PbO = +50,300 \text{ „}$$
$$PbO + H_2SO_4Aq = PbSO_4 + H_2O = +23,400 \text{ „}$$

$$\text{Sum} = +5,300 \text{ cals.}$$

II. At the kathode.

$$PbO_2 + H_2 = PbO + H_2O = +50,300 + 68,400 - (50,300 + 12,140)$$
$$= +56,260 \text{ cals.}$$
$$PbO + H_2SO_4Aq = PbSO_4 + H_2O = +23,400 \text{ „}$$

$$\text{Sum} = +79,660 \text{ cals.}$$

The total thermal value is therefore

$$Q = 5,300 + 79,660 = 84,960 \text{ cals.}$$

From this is given, at 15°,

$$\pi = \frac{84,960}{2 \times 23,090} + 288 \times 0.00032 = 1.975 \text{ volts.}$$

This theoretical value agrees so well with the observed value that the agreement affords the best confirmation of the theory of the chemical changes in accumulators, developed in the preceding chapter, according to which the lead plate and the lead peroxide plate are changed to lead sulphate when the accumulator is discharged.

Jahn has shown, from his investigations, that the secondary heats of reversible cells are to be traced to the *Peltier's effect*. It is known that when a current flows from one metal to another a thermal difference is found at the point of junction

CHAP. IX.] THE ENERGETICS OF GALVANIC ELEMENTS. 217

of the two metals, and at that point only, and that this difference changes its sign according to the direction of the current. This heating, or cooling, effect is called, from the name of its discoverer, *Peltier's effect*. It is proportional to the first power of the intensity of the current, but it also depends, in a way not yet understood, on the substantive nature of the conductor. This effect is most clearly shown with a rod made of antimony and bismuth, when the rod is sufficiently thick, and the current is not too strong, so that the total heat, Q [as previously defined, p. 212], remains small. In figure 54 K is a globe of 4 centims. diameter; r is a tube, added to the globe, which is connected, by the caoutchouc tube s, with one of the manometers of the thermoscope represented in figure 53 (p. 209). The tubuluses t_1 and t_2 are closed by well-fitting corks, through which passes a rod, 8 mm. thick, made partly of antimony, Sb, and partly of bismuth, Bi, the two metals being melted together at l. The ends of the rod are connected with a battery of two accumulators, in the circuit of which a commutator is included, and also, if necessary, a resistance of 1 or 2 ohms. If the current passes from antimony to bismuth, the liquid in the manometer rises about 40 mm. in a few minutes, which signifies that the point of junction,

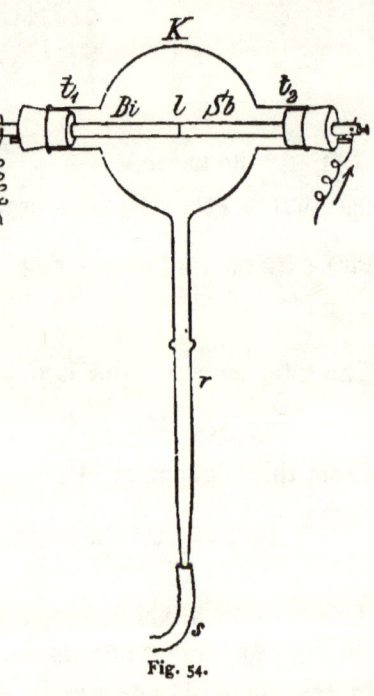

Fig. 54.

l, has become heated. If the current is now reversed, the liquid in the manometer sinks 20 to 30 mm. below its original level, in consequence of the cooling of l; and when the current is stopped the liquid gradually rises to its original level, because of the withdrawal of heat by the point of junction from its surroundings.

The Peltier's effect is clearly observable when a current passes between a conductor of the first class and an electrolyte, but it is small at the surface of separation of two electrolytes. Jahn has measured the Peltier's effect at the points of contact of metals and solutions of their salts. It follows from his data that *the secondary heat of a galvanic element agrees with the algebraic sum of the Peltier's effects that occur at the electrodes.* Hence the secondary heat and the Peltier's effect are probably identical; a conclusion which also follows from the fact that *the secondary heat changes its sign when a current passes into the galvanic element in the reverse direction.*

Finally, it is of great interest to note that Ostwald (*Lehrbuch der allgemeinen Chemie*, vol. ii., p. 858 [1893]) arrived at Helmholtz's equation by starting from the formula

$$\pi = \frac{0.002}{n} \times T . \log \frac{P_1}{P_2}$$

and applying the laws of osmotic pressure to the quantities P_1 and P_2. This result may, therefore, also be regarded as a confirmation of Nernst's theory of the production of the current in galvanic elements.

INDEX.

A.

Accumulators, 4 *note*, 197.
—— capacity of, 203.
—— charging with copper elements, 201.
—— chemical changes in, 198.
—— self-discharging, 205.
—— zero-effect of, 204.
Activity-coefficient, 69.
Alkali salts, electrolysis of, 17.
Aluminium-potassium chloride, 6.
Aluminium salts, electrolysis of, 7.
Amalgam-cells, 142.
Ammonia, electrolysis of, 24.
Ammonium chloride, electrolysis of, 24.
Ampère, definition of, 35 *note*.
Analysis by electrolysis, 183.
Aniline black, electrical preparation of, 31.
Anion, definition of, 3.
Anode, definition of, 3.
Avogadro's law applied to solutions, 116.

B.

Bleaching, electrical, 12.
Boiling points of electrolytes, 118.

Boiling point of a solution, 100.
—— —— molecular raising of, 104.
Brass, electrical formation of, 183.
Bunsen cells, 191.

C.

Calcium carbide, 8.
Calomel cell, 135.
Carborundum, 8.
Cells, amalgam, 142.
—— Bunsen, 191.
—— concentration, 128.
—— Daniell, 136, 141.
—— gas, 152.
—— inconstant, 188.
—— irreversible, 186.
—— Leclanché, 193.
—— liquid, 123.
—— reduction and oxidation, 146.
Chemical energy, change of, into electrical, 147, 151, 212.
—— —— in reduction- and oxidation-cells, 147.
—— —— relation of, to electrical energy, 211.
—— heat of galvanic elements, 215.
Chloride of nitrogen, preparation of, 24.

Chloride of sodium, electrolysis of, 10, 12.
Chlorine ions, velocities of migration of, 51.
Chromic acid cells, 191.
—— —— electrolysis of, 191.
Clark-element, 215 *note*.
Compensation methods, 137.
Complex ions, 24, 62.
Concentration cells, 128.
—— changes of, during electrolysis, 42.
Conduction of electricity in electrolytes, 39, 58, 66, 129.
Conductivity, additive character of molecular, 50.
—— dependence of, on temperature, 52.
—— in electrolytes, 39, 58, 66, 129.
—— Kohlrausch's law of, 50.
—— molecular, 47, 49.
—— Ostwald's law of, 51.
—— specific electrolytic, 45.
—— with increasing dilution, 49.
Conductors of the first class, 5.
—— of the second class, 5.
Constant cells, 188.
Copper element, 201.
—— extraction of, 184.
—— refining of, 183.
—— sulphate, electrolysis of, 14, 41.
Coulomb, definition of, 35 *note*.
Current, conduction of, in electrolytes, 39, 58, 66, 129.
—— division of, 36.

D.

Daniell cells, 136, 141.
Decomposition-tensions, 179.
Depolarisation, 191.
Dissociation coefficients, 59, 69.

Dissociation theory of Arrhenius, 57, 60.
—— of water, 60.
Dropping electrode, 158.

E.

Electro-chemical equivalent, 35.
—— theory of von Helmholtz, 37.
Electrode, definition of, 3.
—— discharging, 129.
—— solution, 129.
Electrolysis, definition of, 3.
—— Grotthus's theory of, 66.
—— phenomena of, 4, 24.
Electrolyte, definition of, 3, 22.
Electrolytes, dissociation of aqueous solutions of, 117.
Electrolytic dissociation into ions, 57, 70, 121.
—— estimation of metals in mixtures, 183.
—— solution-pressures of the metals, 129, 157.
Electromotive force, 137. (See also *Potential-difference*.)
—— —— origin of, in a cell, 164.
—— —— dependence on temperature, 213.
—— series, 165.
Electroplating, 15, 26.
Energetics of galvanic elements, 207.
—— of solutions, 89.
Energy, chemical, change of, into electrical energy, 147, 151, 212.
—— of current, change of, into chemical energy, 211.
—— —— change of, into heat, 58, 211.
—— contents of ions, 65, 68.
Eosin, dissociation of, 61.

Etching, electrical, 15, 16.
Ethereal salts, saponification of, 69.

F.

Faraday's law, 32.
Fixation of ions, intensity of, 171, 176, 181.
Freezing points of electrolytes, 118.
—— —— of solutions, 107.
—— —— molecular lowering of, 109.

G.

Galvanic current, comparison of, with conduction of water, 143.
Gas-accumulators, 154.
Gas-cells, 152.
Gaseous equation, 84.
Gold, electrical extraction of, 26.

H.

Helmholtz's energy equation, 213.
Hittorf's transport-numbers, 41.
Hydrochloric acid, electrolysis of, 11.
Hydrogen ions, charges on, 39.
—— —— velocity of migration of, 56.
—— peroxide, electrical production of, 23.
—— position of, in electromotive series, 166.
Hydroxyl ions, velocity of migration of, 51.

I.

Inconstant cells, 188.
Inlaying metals, 16.

Intensity of current, measurement of, 35.
—— of fixation of ions, 171, 176, 181.
Iodoform, electrical preparation of, 30.
Ionisation, heat of, 63.
—— coefficient, 59 *note*, 69.
Ions, complex, 24, 62.
—— conception of, 3.
—— energy-contents of, 65, 68.
—— migration-velocities of, 51, 53.
—— proof of free, 71.
Iron, galvanised, 170.
—— tinned, 169.

K.

Kanarin, electrical preparation of, 30.
Kathode, definition of, 3.
Kation, definition of, 3.
Kohlrausch, law of, 45.

L.

Lead chloride, electrolysis of, 8.
—— solutions, electrolysis of, 25.
Leclanché cells, 193.
Liquid cells, 123.
Litre-atmosphere, definition of, 85.

M.

Magnesium, electrical preparation of, 4.
—— potassium chloride, preparation of, 4.
Mercurous chloride cell, 135.
Mercury unit, 46.

Metals, inlaying of, 16.
—— colouring of, 25.
—— electrolytic estimation of, in mixtures, 183.
Methylene blue, dissociation of, 62.
Migrations of ions, velocities of, 51, 53.
Molecular weights, determination of, by boiling-point method, 105.
—— —— —— by freezing-point method, 109.
—— —— —— by vapour-pressure method, 96.

N.

Neutralisation, heat of, 71.
Nitration, rate of, 50.
Nitric acid, electrolysis of, 191.
Normal element, 215.

O.

Occlusion by lead, 199.
—— by palladium, 151, 166.
Ohm, definition of, 35 *note*.
Organic compounds, electrical preparation of, 29.
Osmotic work of solutions, 90.
—— pressure, analogy between, and gaseous pressure, 89.
—— —— connection between, and vapour-pressure, 97.
—— —— definition of, 76.
—— —— demonstration of, experimentally, 77.
—— —— magnitude of, 77, 91.
—— —— measurements of, 78, 80.
—— —— Pfeffer's laws of, 84.
—— —— van't Hoff's law of, 85, 88.

Osmotic theory of the formation of the current, 122—218.
Oxidation- and reduction-cells, 146.
Oxygen, preparation of, by electrolysis, 16.
Ozone, formation of, by electrolysis, 23.

P.

Palladium, occlusion of hydrogen by, 151, 166.
Peltier's effect, 217.
Plasmolysis, 77.
Polarisable electrodes, 173.
Polarisation, 41, 191.
Poles, positive and negative, 18.
—— reagent for detecting, 19.
Positive and negative poles, 18.
—— current, direction of, 129.
Potassium chlorate, electrical preparation of, 13.
—— ferrocyanide, electrolysis of, 27.
—— hydroxide, electrolysis of, 8.
—— ions, velocity of migration of, 51.
—— sulphate, electrolysis of, 18.
—— sulphocyanide, electrolysis of, 29.
Potential-difference between a metal and an electrolyte, 130.
—— of concentration-cells, 128.
—— of Daniell cells, 136, 160, 164, 177.
—— of liquid cells, 123.
—— measurement of, by compensation method, 137.
—— —— by dropping electrode, 159.
Purification, electrical, of organic substances, 60.
—— —— of copper, 183.

R.

Reactivities of dissociated compounds, 69.
Reduction- and oxidation-cells, 146.
Resistance, measurement of, in electrolytes, 45.
Reversible cells, 188.

S.

Salt, Berzelius' definition of, 20.
—— Daniell's definition of, 21.
—— Hittorf's definition of, 22.
Salts, double, 70.
Secondary heat of galvanic elements, 214.
—— processes of electrolysis, 24.
Semipermeable membranes, 77.
Silver ions, velocity of migration of, 56.
Silvering, electrical, 15, 26.
Sodium acetate, electrolysis of, 28.
—— chloride, electrolysis of, 10, 12.
—— ions, velocity of migration of, 56.
Solution-pressures of metals, 129, 157.
Solutions, energetics of, 89.
—— of electrolytes, 117.
—— van't Hoff's theory of, 74.
Stannous chloride, electrolysis of, 148.
Sulphuric acid, electrolysis of, 17, 22.

T.

Temperature-coefficient of galvanic elements, 213.
Thermoneutrality, 70.
Tin on iron, 168.
—— tree, 10, 134.
Transport-numbers, 41.

U.

Units, electrical, 35 *note*, 46.
Unpolarisable electrodes, 172.

V.

Valency-charges of ions, 37.
Vapour-pressure, molecular depression of, 93.
Vapour-pressures of solutions, 92.
—— —— connection between, and osmotic pressure, 97.
—— —— laws of, 93.
—— —— —— deduction by van't Hoff's equation, 97.
—— —— proof of, 94.
Volt, definition of, 35 *note*.
Voltameter, 33.

W.

Water, electrolysis of, 22, 60.
—— molecular conductivity of, 60.
Watt, definition of, 35 *note*.

Z.

Zinc on iron, 170.
—— chloride, electrolysis of, 9.

www.ingramcontent.com/pod-product-compliance
Lightning Source LLC
Chambersburg PA
CBHW021804230426
43669CB00008B/627